Process Control Techniques for High-Volume Production

Process Control Techniques for High-Volume Production

M. Kemal Atesmen

CRC Press
Taylor & Francis Group
Boca Raton London New York

CRC Press is an imprint of the
Taylor & Francis Group, an **informa** business
AN AUERBACH BOOK

CRC Press
Taylor & Francis Group
6000 Broken Sound Parkway NW, Suite 300
Boca Raton, FL 33487-2742

First issued in hardback 2019

First issued in paperback 2022

© 2017 by Taylor & Francis Group, LLC
CRC Press is an imprint of Taylor & Francis Group, an Informa business

No claim to original U.S. Government works

ISBN 13: 978-1-03-247724-4 (pbk)
ISBN 13: 978-1-4987-6911-2 (hbk)

DOI: 10.1201/9781315367224

Library of Congress Cataloging-in-Publication Data

Names: Atesmen, M. Kemal, author.
Title: Process control techniques for high-volume production / M. Kemal Atesmen.
Description: Boca Raton : Taylor & Francis, CRC Press, 2017. | Includes bibliographical references and index.
Identifiers: LCCN 2016006110 | ISBN 9781498769112
Subjects: LCSH: Process control. | Manufacturing processes. | Production management.
Classification: LCC TS156.8 .A775 2017 | DDC 658.5--dc23
LC record available at https://lccn.loc.gov/2016006110

Visit the Taylor & Francis Web site at
http://www.taylorandfrancis.com

and the CRC Press Web site at
http://www.crcpress.com

Contents

CONTENTS

List of Figures

Preface

This book details the most common statistical process control tools for high-volume production. Although several books have been written on this subject, all of the necessary statistical process control books are scattered, are difficult to understand, and do not have constructive examples. In this book, my aim is to make elements of high-volume production process control simpler and easier to understand. Many software products have also been developed on this subject. However, they are like secret boxes. One inputs data blindly and gets results out without understanding the actual statistical process control thought process. Anyone dealing with high-volume production as an operator, line supervisor, inspector, process engineer, quality engineer, manufacturing manager, plant manager, all the way up to the president of the company needs to understand and absorb the following statistical process control basics that are outlined in this book to achieve success.

Manufacturing a product such as a car with more than 10,000 components fitting together and functioning properly when completed in a high volume requires very well defined process controls. A manufacturing plant in high-volume production should have strict procedures in place as to what to do and how to deal with out-of-specification parts and out-of-control process conditions, including shutting down production and calling emergency meetings. The people who should be involved all the way to top management of the company should be clearly identified for every process step. Building a product in high volume starts with an accurate and detailed set of technical drawings and specifications. These technical drawings and specifications determine critical parameters and measurement standards that have to be used in high-volume production.

Dealing with product specifications, critical parameters, and measurement standards is discussed in Chapter 1. Measurements are done with gages that have to be capable in the short term and in the long term. Short-term gage capability evaluation for variables is detailed in Chapter 2 and long-term gage capability evaluation for variables in Chapter 3. The most vulnerable portion of high-volume production is inspection stations for attributes that need utmost attention and discipline. Gage capability for attribute data is discussed in Chapter 4.

Once product specifications, critical parameters, measurement standards, and capable gages are in place, statistical process controls for variables and for attributes are detailed. During high-volume production control, distribution of product sampling averages for a variable can be approximated by a normal distribution. Also, the mean of the product sampling averages for a variable equals the total product mean. In addition, most variables data in high-volume production can be approximated by a normal distribution. These statistical characteristics during high-volume production generate a need to understand normal distribution well, which is detailed in Chapter 5.

The most popular and useful statistical process control tools for variables in high-volume production are $\bar{X}$ and R control charts, detailed in Chapter 6, and $\bar{X}$ and S control charts, detailed in Chapter 7. The most popular and useful statistical process control tools for attributes in high-volume production are P, NP, C, and U control charts, detailed in Chapter 8.

Another very important segment of high-volume production is correlation between gages on our production lines, with our customers' gages and with our subcontractors' gages. Correlation between two gages is discussed in Chapter 9.

For every specification and critical parameter of our product, high-volume production processes have to be capable to 6σ or better–see Chapter 10 for 6σ process capability definition–in order not to rely on expensive and time consuming inspections of 100% sorting.

Acknowledgments

More than 33 years of engineering project and process management in the global high-volume production arena covering automotive, computer, data communication, and offshore oil industries were accomplished by exceptional support from my wife, Zeynep, and my family members. On some occasions I was away from home more than six months out of a year trying to tackle challenging project tasks.

I would like to dedicate this book to all project team and production members whom I had the pleasure of working with over the years, who did the hard work with enthusiasm, and who kept coming back to work along with me on a project team and in a high-volume production environment without any reservations.

Introduction

Statistical process control tools in high-volume production can be very complex, difficult to understand, and applied mostly blindly on production lines without a full comprehension as to what is going on. This book details the most common statistical process control tools with examples that were encountered by the author in a high-volume production environment.

Once we have process specifications and critical parameters clearly defined and understood, traceability of process measurements to international standards is at the top of the list for high-volume production control. Product specifications, critical parameters, measurement standards, and gages in high-volume production control are discussed in Chapter 1. Hierarchy of measurement standards is described. Examples of gages that are inaccurate but precise, accurate but imprecise, and accurate and precise are given.

Measurements are done with gages that have to be capable in the short term and in the long term. Step-by-step short-term gage capability evaluation for variables is detailed in Chapter 2 through examples. In one example for short-term gage repeatability and reproducibility, one gage and multiple test operators are used. Gage capability is improved by retraining one of the test operators. In another short-term gage capability evaluation example for variables, one gage and one experienced test operator are used. In this example, gage capability is improved by deleting unstable test parts.

Long-term gage capability evaluation for variables is detailed in Chapter 3. In one example a high quantity of monitoring products is used to determine long-term gage capability of a digital multimeter. In another example a low quantity of monitoring products is used utilizing different control charts to determine long-term gage capability. How to treat out-of-control conditions in gage monitoring control charts is also discussed.

The most vulnerable portion of high-volume production is inspection stations for attributes that need utmost attention and discipline. Gage capability for attribute data is discussed in Chapter 4. Repeated inspection evaluations by several inspectors using multiple test products are analyzed. The repeatability and reproducibility for each inspector are detailed step by step. In another example an automated laser measurement system with go and no-go outputs is evaluated for long-term gage capability.

Once product specifications, critical parameters, measurement standards, and capable gages are in place, statistical process controls for variables and for attributes are detailed during the high-volume production process. During high-volume production process control, distribution of product sampling averages for a variable can be approximated by a normal distribution. Also, the mean of the product sampling averages for a variable equals the total product mean. In addition, most variables data in high-volume production can be approximated by a normal distribution. These statistical characteristics during high-volume production generate a need to understand normal distribution well, which is detailed in Chapter 5. Characteristics of the normal distribution function are detailed. Examples of areas under the normal distribution function, namely cumulative probabilities, are calculated using an MS Excel® function. These calculations led to estimates of total number of defective products for a given variable in high-volume production.

The most popular and useful statistical process control tools for variables in high-volume production are $\bar{X}$ and R control charts, detailed in Chapter 6, and $\bar{X}$ and S control charts, detailed in Chapter 7. $\bar{X}$ and R control charts are described in Chapter 6 by means of examples. Calculations of upper and lower control limits are shown step by step. Out-of-control conditions are discussed by analyzing examples. An example is provided to show how to achieve a 6σ process capability. $\bar{X}$ and S control charts are detailed in Chapter 7 by means of examples. Calculations for upper and lower control limits in $\bar{X}$ and S control charts are also shown step by step. Statistically out-of-control conditions are discussed by analyzing several examples. An example of process improvement verification is shown through $\bar{X}$ and S control charting.

The most popular and useful statistical process control tools for attributes in high-volume production are P, NP, C, and U control charts, detailed in Chapter 8. When to use these different attribute control charts is discussed and examples are provided for every case. Calculations of upper and lower control limits and evaluation for out-of-control conditions for different types of attribute control charts are detailed. The advantages of using constant sample sizes versus more cumbersome and confusing varying sample sizes while attribute control charting are discussed.

Another very important segment of high-volume production process control is the correlation between gages on our production lines with our customers' gages and with our subcontractors' gages. Correlation between two gages is discussed in Chapter 9. Correlations between gages with linear and nonlinear behavior and with positive and negative slopes are detailed. Cases exhibiting strong, weak, and no correlation are given.

For every specification and critical parameter of our product, high-volume production processes have to be capable to 6σ or better–see Chapter 10 for 6σ process capability definition–in order not to rely on expensive and time consuming inspections of 100% sorting. Process capability analysis for variables is detailed in Chapter 10. Once our high-volume production processes are in statistical control and are normally distributed, process capability for a variable can be obtained by using an MS Excel function. Several examples of process capability calculations and calculations for the number of products that are out of specifications are provided. Also, in a case example, necessary process improvements are detailed to meet our customer's 6σ requirement.

About the Author

M. Kemal Atesmen completed his high school studies at Robert Academy in Istanbul, Turkey, in 1961. He received his BSc degree from Case Western Reserve University, his MSc degree from Stanford University, and his PhD degree from Colorado State University, all in mechanical engineering. He is a life member of ASME. He initially pursued an academic and an industrial career in parallel and became an associate professor in mechanical engineering before dedicating his professional life to international engineering project management and engineering management for 33 years. He helped many young engineers in the international arena to bridge the gap between college and professional life in automotive, computer component, data communication, and offshore oil industries.

He is the author of five books: *Global Engineering Project Management* (CRC Press, 2008), *Everyday Heat Transfer Problems: Sensitivities to Governing Variable*s (ASME Press, 2009), *Understanding the World Around through Simple Mathematics* (Infinity Publishing, 2011), *A Journey Through Life* (Wilson Printing, 2013), and *Project Management Case Studies and Lessons Learned: Stakeholder, Scope, Knowledge, Schedule, Resource and Team Management* (CRC Press, 2014). In addition, he has published 16 technical papers and holds four patents.

1

PRODUCT SPECIFICATIONS, CRITICAL PARAMETERS, AND MEASUREMENT STANDARDS

Building a product starts with an accurate and detailed set of technical drawings and specifications. A final assembled product can have several subassemblies and components. Each level of assembly has critical dimensions and parameters, all of which have specification limits. An engineering team has to go through several steps and negotiations to generate these critical dimensions and parameters. After many meetings with customers, manufacturing groups, and quality groups, final values of critical dimensions and parameters and their tolerances are settled.

An agreed-upon tolerance depends on the manufacturing process capability, appropriate level of measurement resolution, and the measurement's gage capability. Before we can discuss and build our manufacturing structure, we have to agree on international or other standards that will be traceable from our measurements. Even the international standards evolve to higher accuracy as dimensions and tolerances get smaller and smaller. Accuracy is defined as the deviation of the measured value from the true one. For example, the unit of length, the meter, evolved from the length of a prototype platinum–iridium bar to the wavelength of light that travels in a vacuum during a time interval of 1/299,792,458 of a second. However, the standard for mass, namely one kilogram of mass, is a platinum–iridium object made in the nineteenth century and is kept under specific atmospheric and contaminant-free conditions. Also, the unit of time, the second, evolved from a fraction of the mean solar day, namely 1/(24 × 3600), which had irregularities due to Earth's rotation to a transition between two energy levels of a cesium-133 atom at a frequency of 9,192,631,770 Hertz. As industrialization evolved, similar standards were established for unit of electric current, namely ampere, for unit of thermodynamic temperature, namely Kelvin, for unit of amount of substance, namely mole and for unit of luminous intensity, namely candela.

Production line working standards have to be traceable to silver standards in our company's divisions and at our customers' facilities. Then all silver standards have to be traceable to our company's gold standard, which in turn has to be traceable to international standards through the American National Standards Institute. All of these standards have to be kept in protective cases and sometimes in specified environments. These standards represent true values of our critical dimensions and parameters. For example, a 100-nm glass layer thickness standard between two ferrite surfaces for deposition time calibration, a 1-megaohm resistor standard with excellent temperature stability for online digital multimeter calibration, and a 1-g mass standard with a ±0.005 g tolerance to true value and traceable to the international mass standard are examples of gold standards in a company's calibration laboratory. These gold standards are obtained from accredited metrology laboratories. They are used to calibrate a well-preserved gage in the company's calibration laboratory. Silver and working standards are then generated using this calibrated gage. The hierarchy of measurement standards is summarized in Table 1.1. Silver and working standards are also protected well from environmental effects. Once a production line gage is calibrated, its short- and long-term capabilities are realized by using monitoring product samples.

Table 1.1   Hierarchy of Standards from the International Base All the Way Down to a Production Line

MEASUREMENT STANDARD TYPE	MEASUREMENT STANDARD KEEPER
International standards	General conference on weights and measures
Traceable U.S. standards	American National Standards Institute
Traceable standards generators	Accredited metrology laboratories
Traceable gold standard	Your company's calibration laboratory
Traceable silver standards	Your company's calibration laboratory, your customers, and your subcontractors
Traceable working standards	Production lines
Product samples for gage monitoring	Production lines

Before we can talk about process capability in our production, we have to determine how capable our gage is for measuring that certain critical dimension or parameter. Our gage has to be accurate and precise to be able to assess our process capability. Precision of a gage is defined as the deviation of a group of repeated measurements from a mean value. A gage can be accurate but not precise, or it can be precise but not accurate. The best gages should be both accurate and precise.

One way to visualize the effects of gages on measurements and on our products is to go through several examples. Let us consider the measurement of a critical length specification of 28.0 ± 6.0 μm for our product. After measuring this critical length on thousands of our products using an accurate and precise gage, we obtained an average of 27.0 μm and a standard deviation—spread—of 2.1 μm. This critical length dimension behaved like a normal distribution; see Chapter 5 for normal distribution details. During these measurements, our product was produced through stable and statistically controlled production processes.

We then introduced a new length measurement gage to our production line and measured repeatedly a production line standard with a true value of 28.0 μm and obtained the results shown in Figure 1.1. This new gage was inaccurate because the average reading was 1.8 μm lower than the true value of the standard. However, this gage was precise and it had a standard deviation, σ_{gage}—spread—of

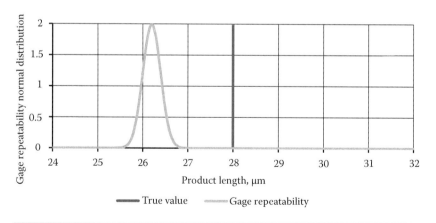

Figure 1.1 Precise and inaccurate gage measurement where the mean of the gage measurements is 1.8 μm lower than the standard's true value and $\sigma_{gage} = 0.2$ μm provides a precision to total tolerance ratio of 10%.

0.2 μm. The precision to total tolerance ratio for this new gage was 0.10; see Chapter 2 for precision to total tolerance ratio details, namely Precision to total tolerance ratio $= \dfrac{6 \times \sigma_{\text{gage}}}{\text{USL} - \text{LSL}} = \dfrac{6 \times 0.2}{34 - 22} = 0.10$ or 10% of total tolerance. USL refers to an upper specification limit of 34.0 μm. LSL refers to a lower specification limit of 22.0 μm.

We introduced a second length measurement gage to our production line and measured repeatedly the same production line standard with a true value of 28.0 μm and obtained the results shown in Figure 1.2. This new gage was accurate because the average reading was 0.15 μm lower than the true value of the standard. However, this second gage was not precise and it had a standard deviation—spread—of 1.0 μm. The precision to total tolerance ratio for this new gage was 0.5 (see Chapter 2), namely Precision to total tolerance ratio $= \dfrac{6 \times \sigma_{\text{gage}}}{\text{USL} - \text{LSL}} = \dfrac{6 \times 1.0}{34 - 22} = 0.5 \text{ or } 50\%$ of total tolerance.

Finally, we found a third gage that was both accurate and precise in measuring this critical length dimension. Repeated measurement results for this third gage are shown in Figure 1.3. The mean of repeated length measurements with our production line standard shifted only 0.10 μm from its true value of 28.0 μm. The spread of the

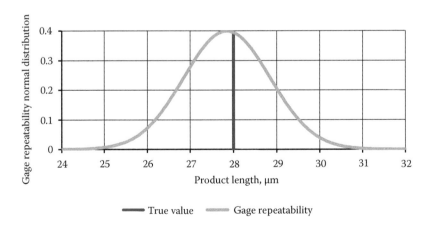

Figure 1.2 Accurate and imprecise gage measurement where the standard deviation—spread—of the second gage measurements was 1.0 μm, which resulted in a precision to total tolerance ratio of 50%. This new gage was accurate because the average reading was 0.15 μm lower than the true value of the standard.

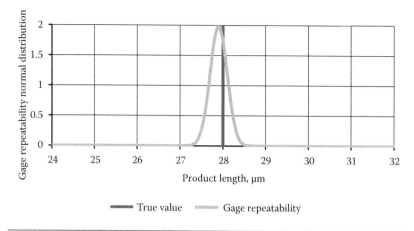

Figure 1.3 An accurate and precise gage. Mean of measurements was 0.10 µm lower than the standard's true value. The spread of measurements was 0.20 µm, which provided us with a precision to total tolerance ratio of 10%.

repeated measurements was only 0.20 µm, which provided a precision to total tolerance ratio of 10%.

After measuring this critical length dimension with an accurate and precise gage on many randomly chosen products throughout several months of stable production, we obtained the process normal distribution given in Figure 1.4. Our process had an overall average of 27.0 µm and a process standard deviation—spread—of 2.1 µm. Our processes for this critical length were not capable; see Chapter 10 for process capability calculations. We were rejecting 0.86% of our

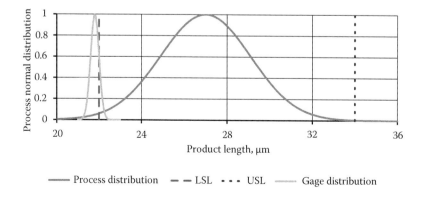

Figure 1.4 Even an accurate and precise gage can make a bad part a good one.

product below our lower specification limit of 22.0 μm and 0.043% of our product above our upper specification limit of 34.0 μm. Therefore we had to sort our product 100% for this critical length dimension. However, we had a dilemma. Even if we measured this critical length with an accurate and capable gage, we could accept substandard parts that had lengths lower than our LSL as shown in Figure 1.4. We had a 16% chance of accepting a reject product that had a length of 21.8 μm. Therefore for products whose lengths were close to upper and lower specification limits, we had to perform multiple length measurements to obtain their true values. Another typical case for using an accurate and precise gage to accept and to reject products that are right on the lower and on the upper specification limits are shown in Figure 1.5.

As we can see from the preceding examples, measurement gages on our production lines are very crucial decision making centers for producing specification-compliant products. Production line measurement gages are calibrated by the company's calibration laboratory personnel, and their calibrations should always be traceable to international standards. Stickers are affixed on the gage telling us the calibration date, who calibrated it, and the calibration expiration date. If a gage is ready to be used on the production line, we first have to perform a short-term gage capability evaluation using product samples for a particular critical dimension or parameter. We want to make sure that the gage we are going to use to measure a particular critical dimension or parameter is

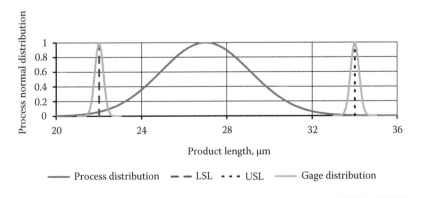

Figure 1.5 An accurate and precise gage has a 50–50 chance of calling a good or a bad product when the true value of a good product is right on LSL or on USL.

capable within given tolerances of finding the true value of our products. The measurements we make on our production line have to be traceable to national and to international standards, which is a requirement for International Organization for Standardization (ISO) certification.

2

Short-Term Gage Capability Evaluation for Variables

First we have to establish that our measurement gage is in control and capable—accurate and precise—in the short term, namely in a day or at most in a week. After establishing short-term gage capability, we can extend our gage capability investigation to the long term, namely to a month or more. After verifying that our measurement gage is in control and capable in the long term, we need to keep monitoring our gage with our production line standards in fixed periods, that is, hourly, daily, or weekly, and control the chart standards' measurements to verify continuously our gage's in-control status.

In this chapter, I provide two examples of short-term gage capability analysis. The first example details a gage repeatability and reproducibility analysis using one gage and three test operators. The second example details a gage repeatability analysis using only one gage and one certified and experienced test operator.

The following steps should be followed for any short-term gage capability evaluation procedure for variables:

- Products chosen for gage capability should be chosen at random and should represent the true variation of our production process.
- All measurements should be done in a short time period, that is, an hour, a day, or at most a week, and all measurements should be in stable environmental conditions.
- All test operators who will be using the gage should be trained and certified to use that particular gage. We need at least two test operators for short-term gage reproducibility analysis.
- Each test operator should measure each product at least twice at random and record his or her results in a short time period

and under the same environmental conditions to evaluate gage repeatability.

The following short-term gage capability evaluation is for measuring the resistance of a coil by a digital multimeter with serial number 90128. We used three test operators and 10 coils selected at random. Each test operator repeated each coil measurement twice. The short-term gage capability data sheet was filled out as shown in Table 2.1.

Now that we have the measurement data let us go through gage capability evaluation calculations step by step.

STEP 1: Calculate ranges of measurements for every product and for every test operator, averages of ranges for every test operator, and averages of all product measurements for each test operator.

The gage capability evaluation for the data presented in Table 2.1 started with the measurement data sheet in the table. The range for each product's two measurements was calculated in the third column for that test operator. The range was simply the maximum reading for that product minus the minimum reading for the same product.

The average of the measurement ranges for each test operator is obtained for 10 sample products. These averages are also calculated on the measurement data sheet in Table 2.1 and also shown for test operator 1 in Equation 2.1.

The average of measurement ranges for test operator 1 is obtained from 10 sample products—each measured twice—as follows:

$$(0.022 + 0.009 + 0.002 + 0.012 + 0.012 + 0.031 + 0.018 + 0.018 + 0.011 + 0.015)/10 = 0.015 \tag{2.1}$$

The average of all 20 measurements for each test operator is also shown in Table 2.1. This average calculation for all 20 measurements by test operator 1 is shown as an example in Equation 2.2.

The average of 20 measurements for test operator 1 is obtained from 10 sample products as follows:

$$(5.222 + 5.244 + 4.901 + 4.910 + 5.383 + 5.381 + 5.290 + 5.302 + 5.515 + 5.503 + 5.181 + 5.212 + 5.314 + 5.332 + 5.112 + 5.130 + 5.284 + 5.273 + 5.247 + 5.262)/20 = 5.2499 \tag{2.2}$$

STEP 2: Calculate the average range for all test operators and gage repeatability variance.

Table 2.1 Sample Short-Term Gage Capability Measurement Data Sheet for a Variable

GAGE: Digital Multimeter, Serial No. 90128
PART NAME: Wound Coil
CRITICAL DIMENSION: Full Coil Resistance
DATE: September 3, 2004
PART NUMBER: 337901
TOLERANCE: 5.0 ± 0.5 Ohms

PRODUCT	OPERATOR 1: ALICE			OPERATOR 2: BENJAMIN			OPERATOR 3: WALTER		
	FIRST TRIAL	SECOND TRIAL	ABSOLUTE RANGE	FIRST TRIAL	SECOND TRIAL	ABSOLUTE RANGE	FIRST TRIAL	SECOND TRIAL	ABSOLUTE RANGE
P1	5.222	5.244	0.022	5.116	5.190	0.074	5.235	5.224	0.011
P2	4.901	4.910	0.009	4.852	4.804	0.048	4.932	4.950	0.018
P3	5.383	5.381	0.002	5.513	5.562	0.049	5.364	5.381	0.017
P4	5.290	5.302	0.012	5.415	5.551	0.136	5.247	5.222	0.025
P5	5.515	5.503	0.012	5.530	5.487	0.043	5.493	5.528	0.035
P6	5.181	5.212	0.031	5.321	5.363	0.042	5.214	5.183	0.031
P7	5.314	5.332	0.018	5.312	5.464	0.152	5.294	5.301	0.007
P8	5.112	5.130	0.018	5.147	5.121	0.026	5.101	5.105	0.004
P9	5.284	5.273	0.011	5.238	5.256	0.018	5.250	5.241	0.009
P10	5.247	5.262	0.015	5.304	5.372	0.068	5.252	5.237	0.015
Averages of 10 ranges	Operator 1 =		0.015	Operator 2 =		0.0656	Operator 3 =		0.0172
Averages of 20 measurements	Operator 1 =		5.2499	Operator 2 =		5.2959	Operator 3 =		5.2377

We now have to calculate the average range for all three operators as shown in Equation 2.3. From this average range, we can obtain the gage repeatability variance by dividing the average range by a factor as shown in Equations 2.4 and 2.5.

Average of ranges for all three test operators = (0.0150 + 0.0656 + 0.0172)/3 = 0.0326 (2.3)

Gage repeatability variance = [(Average of ranges for all three operators)/(Factor for converting average range into gage repeatability variance, d_2^*)]^2 (2.4)

The factor for converting average range into gage repeatability variance, d_2^*, is obtained from A. J. Duncan's *Quality Control and Industrial Statistics* (1974), p. 950, Table D3, or Table 2.2 in this chapter. For this case, namely for a total of 30 range measurements and 2 measurement values for each range, the conversion factor is 1.13.

Gage repeatability variance = (0.0326/1.13)^2 = 0.000832 (2.5)

The factor for converting average range into gage repeatability variance, d_2^*, is also presented in Table 2.2 for a different number of repeated measurements—2 in this case—and for a different number of sample ranges used—overall 30 sample ranges in this case.

STEP 3: Calculate the range of all operators' measurement averages and test operator reproducibility variance.

For operator reproducibility variance, we use the range of three operators' measurement averages. Operators used the same gage and measured the same 10 parts at random in a short time interval under the same environmental conditions. First we need to calculate the range of three operators' measurement averages as shown in Equation 2.6.

Range of operators' measurement averages:
(Maximum operator average – Minimum operator average)
= 5.2959 – 5.2377 = 0.0582 (2.6)

From the range of three operators' measurement averages, we can obtain the gage reproducibility variance by dividing the range calculated in Equation 2.6 by a different factor, d_2^*, as shown in Equations 2.7 and 2.8.

Table 2.2   Factor, d_2^*, for Converting Measurement Range into Variance from A. J. Duncan's *Quality Control and Industrial Statistics* (1974), p. 950, Table D3

		NUMBER OF REPEATED MEASUREMENTS FOR EACH SAMPLE PRODUCT								
		2	3	4	5	6	7	8	9	10
Number of sample products used in gage capability analysis	1	1.41	1.91	2.24	2.48	2.67	2.83	2.96	3.08	3.18
	2	1.28	1.81	2.15	2.40	2.60	2.77	2.91	3.02	3.13
	3	1.23	1.77	2.12	2.38	2.58	2.75	2.89	3.01	3.11
	4	1.21	1.75	2.11	2.37	2.57	2.74	2.88	3.00	3.10
	5	1.19	1.74	2.10	2.36	2.56	2.73	2.87	2.99	3.10
	6	1.18	1.73	2.09	2.35	2.56	2.73	2.87	2.99	3.10
	7	1.17	1.73	2.09	2.35	2.55	2.72	2.87	2.99	3.10
	8	1.17	1.72	2.08	2.35	2.55	2.72	2.87	2.98	3.09
	9	1.16	1.72	2.08	2.34	2.55	2.72	2.86	2.98	3.09
	10	1.16	1.72	2.08	2.34	2.55	2.72	2.86	2.98	3.09
	11	1.16	1.71	2.08	2.34	2.55	2.72	2.86	2.98	3.09
	12	1.15	1.71	2.07	2.34	2.55	2.72	2.85	2.98	3.09
	13	1.15	1.71	2.07	2.34	2.55	2.71	2.85	2.98	3.09
	14	1.15	1.71	2.07	2.34	2.54	2.71	2.85	2.98	3.08
	15	1.15	1.71	2.07	2.34	2.54	2.71	2.85	2.98	3.08
	Greater than 15	1.13	1.69	2.06	2.33	2.53	2.70	2.85	2.97	3.08

Note: Table D3 first appeared as a whole in the *Journal of the American Statistical Association*, Volume 53, No. 282, p. 548 on June 1958 in his own article entitled "Design and operation of a double-limit variables sampling plan" published by Taylor & Francis.

Table D3 is reproduced with permission in part from Table 30A of "Biometrika Tables for Statisticians" Volume 1 and in part from Acheson J. Duncan, "The use of ranges in comparing variabilities," *Industrial Quality Control*, Volume XI, No. 5, February 1955. Part of Table D3 has been computed by the author.

Operator reproducibility variance = [(Range of operators' measurement averages)/(Factor for converting range of operators' measurement averages into operator reproducibility variance, d_2^*)]²

$$(2.7)$$

The factor for converting range of operators' measurement averages into operator reproducibility variance, d_2^*, is again obtained from A. J. Duncan's *Quality Control and Industrial Statistics* (1974), p. 950, Table D3, or from Table 2.2 in this chapter. For the present case, we use one measurement parameter—average of measurements for each operator—and the range of three average values. Here the conversion factor is 1.91.

Operator reproducibility variance = $(0.0582/1.91)^2 = 0.000928$
$$(2.8)$$

STEP 4: Calculate the measurement average for each product, the range for these averages, and the product variance.

Next we have to determine the product variance for these 10 samples. The average values for each product, obtained from six measurements for that product, are shown in Table 2.3. The range of 10 products' averages is also given in Table 2.3.

We can now calculate the product variance from the range of 10 products' averages by dividing it by a different factor, d_2^*, as shown in Equations 2.9 and 2.10.

Product variance = [(Range of products' averages)/(Factor for converting range of products' averages into product variance, d_2^*)]^2
$$(2.9)$$

The factor for converting the range of products' averages into product variance, d_2^*, is again obtained from A. J. Duncan's *Quality Control and Industrial Statistics* (1974), p. 950, Table D3, or from Table 2.2 in this chapter. For the present case, we use 1 measurement parameter— average of 6 measurements for each sample product—and range for 10 sample products' average values. Here the conversion factor is 3.18.

Product variance = $(0.617/3.18)^2 = 0.037748$ $\qquad$ (2.10)

Table 2.3 Average Measurement Values (Average of Six Measurements) for Each Product and the Range of 10 Products' Averages in Ohms

PRODUCT NO.	PRODUCT AVERAGE
P1	5.205
P2	4.892
P3	5.431
P4	5.338
P5	5.509
P6	5.246
P7	5.336
P8	5.119
P9	5.257
P10	5.279
Range	0.617

STEP 5: Calculate the total of three variances, namely gage repeatability, gage (test operator) reproducibility and product variability, and the ratio of each variance to the total variance.

The total variance for the short-term gage capability analysis is shown in Equations 2.11 and 2.12.

Total variance for short-term gage capability evaluation = Gage repeatability variance + Operator reproducibility variance + Product variance (2.11)

Total variance for this short-term gage capability evaluation
= 0.000832 + 0.000928 + 0.037748 = 0.039508 (2.12)

Then we can obtain the ratio of each type of variance to the total variance as shown in Equations 2.13 through 2.16.

Gage repeatability variance = (0.000832/0.039508) × 100
= 2.11% of total variance (2.13)

Gage reproducibility variance = 2.35% of total variance (2.14)

Gage repeatability and gage reproducibility variance = (2.11% + 2.35%) = 4.46% of total variance (2.15)

Product variance = 95.54% of total variance (2.16)

As we can see from the above results, the sum of gage repeatability variance and reproducibility variance is only 4.46% of the total variance. Is this gage repeatability variance and reproducibility variance good enough for the precision of our measurement system? We have to analyze this question next.

STEP 6: Calculate the precision of the measurement system (gage plus operators) to total tolerance ratio and decide if you can use this gage and work with these test operators on your production line or not.

The precision of the gage to total tolerance ratio is a good guide to see how our gage and operators are performing in measuring a critical parameter of our product. This ratio is defined as follows:

Precision of gage to total tolerance ratio = 100 × [6 × square root (Gage repeatability variance + Gage reproducibility variance)]/(Total critical dimension tolerance) (2.17)

If the precision of gage to total tolerance ratio is less than 10%, the gage that is being evaluated and test operators are precise enough for measuring this critical parameter of our product. The measurement system as a whole has a good enough precision to measure this critical parameter. If the precision of gage to total tolerance ratio is between 10% and 30%, the gage that is being evaluated and the test operators are providing a marginal precision in measuring this critical parameter of our product. The measurement system as a whole needs improvement and then repeated measurements have to be performed again to show a precision of gage to total tolerance ratio less than 10%. If the precision of gage to total tolerance ratio is greater than 30%, you might be wasting your time trying to improve the measurement system. It might be more advantageous to get a new gage and/or work with new test operators.

$$\text{Precision of gage to total tolerance ratio} = 100$$
$$\times [6 \times \text{square root } (0.000832 + 0.000928)]/(5.5 - 4.5) = 25.2\%$$

$$(2.18)$$

We see that the precision of the gage to total tolerance ratio given in Equation 2.18 for the measurement data in Table 2.1 is higher than 10%. We cannot use this measurement system (gage plus operators) to check this critical parameter for our product.

We should first review the measurement data in Table 2.1. When we observe the average test operators' ranges for three test operators in Table 2.1, we can easily see that test operator 2 has a problem. His measurement range is more than four times that of the ranges of the other two test operators. He definitely needs retraining. If we can improve his measurement techniques and therefore his measurement ranges, then we can definitely improve the precision of our gage to total tolerance ratio.

There are other ways to analyze the short-term gage repeatability and reproducibility data given in Table 2.1. Several different kinds of software are available in the marketplace for analysis of variance.

Test Operator 2, Benjamin, was retrained and he remeasured the same 10 parts twice. We obtained the results in Table 2.4 with a much improved average range value—improved 0.0170 Ohms average range versus the previous 0.0656 Ohms average range—for him.

Table 2.4 Sample Short-Term Gage Capability Measurement Data Sheet for a Variable after Retraining Test Operator 2

GAGE: Digital Multimeter, Serial No. 90128
PART NAME: Wound Coil
CRITICAL DIMENSION: Full Coil Resistance
DATE: September 4, 2004
PART NUMBER: 337901
TOLERANCE: 5.0 ± 0.5 Ohms

PRODUCT	OPERATOR 1: ALICE			OPERATOR 2: BENJAMIN			OPERATOR 3: WALTER		
	FIRST TRIAL	SECOND TRIAL	ABSOLUTE RANGE	FIRST TRIAL	SECOND TRIAL	ABSOLUTE RANGE	FIRST TRIAL	SECOND TRIAL	ABSOLUTE RANGE
P1	5.222	5.244	0.022	5.116	5.121	0.005	5.235	5.224	0.011
P2	4.901	4.910	0.009	4.852	4.884	0.032	4.932	4.950	0.018
P3	5.383	5.381	0.002	5.402	5.390	0.012	5.364	5.381	0.017
P4	5.290	5.302	0.012	5.245	5.211	0.034	5.247	5.222	0.025
P5	5.515	5.503	0.012	5.532	5.503	0.029	5.493	5.528	0.035
P6	5.181	5.212	0.031	5.329	5.331	0.002	5.214	5.183	0.031
P7	5.314	5.332	0.018	5.314	5.303	0.011	5.294	5.301	0.007
P8	5.112	5.130	0.018	5.142	5.161	0.019	5.101	5.105	0.004
P9	5.284	5.273	0.011	5.236	5.251	0.015	5.250	5.241	0.009
P10	5.247	5.262	0.015	5.304	5.315	0.011	5.252	5.237	0.015
Averages of 10 ranges	Operator 1 =	0.0150		Operator 2 =	0.0170		Operator 3 =	0.0172	
Averages of 20 measurements	Operator 1 =	5.2499		Operator 2 =	5.2471		Operator 3 =	5.2377	

The results in Table 2.4 can be reevaluated similarly to previous calculations, Equations 2.1 through 2.18, for short-term gage capability evaluation.

STEP 1: Calculate the ranges of measurements for every product and for every test operator, averages of ranges for every test operator, and averages of all product measurements for each test operator.

Gage capability evaluation calculations for the data presented in Table 2.4 started with the measurement data sheet in the table by calculating two repeated measurement ranges for every product and for every test operator, by calculating the average of 10 products' measurement ranges for each test operator, and by calculating the average of 20 product measurements for each test operator.

STEP 2: Calculate the average range for all test operators and gage repeatability variance.

Next we have to calculate the average range for all three test operators as shown in Equation 2.19. From this average range, we can obtain the improved gage repeatability variance by dividing the average range by a factor, d_2^*, as shown in Equation 2.20. For this case, namely for a total of 30 range measurements and 2 measurement values for each range, the conversion factor is 1.13.

$$\text{Average of ranges for all three operators} = (0.0150 + 0.0170 + 0.0172)/3 = 0.0164 \tag{2.19}$$

Improved gage repeatability variance becomes

$$\text{Improved gage repeatability variance} = (0.0164/1.13)^2 = 0.000211 \tag{2.20}$$

STEP 3: Calculate the range of all operators' measurement averages and test operator reproducibility variance.

From the range of three test operators' measurement averages, Equation 2.21, we can obtain the improved test operators' reproducibility variance in Equation 2.22. For the conversion factor, d_2^*, we use one measurement parameter—average of measurements for each operator—and the range of three average values. The conversion factor is therefore 1.91.

The range of operators' measurement averages changes to
(Maximum operator average − Minimum operator average)

$$= 5.2499 - 5.2377 = 0.0122 \qquad (2.21)$$

The improved test operator reproducibility variance is given in Equation 2.22.

Improved operator reproducibility variance = $(0.0122/1.91)^2$
= 0.000041 $\qquad\qquad\qquad\qquad\qquad\qquad\qquad$ (2.22)

STEP 4: Calculate the measurement averages for each product, the range for these averages, and the product variance.

Next we have to determine the product variance in these 10 samples with the new measurements for the second test operator. The average values for each product, obtained from six measurements for that product, are shown in Table 2.5. The new product variance is calculated in Equation 2.23. For the conversion factor, d_2^*, we use 1 measurement parameter—the average of six measurements for each sample product—and the range for 10 sample products' average values. In this case the conversion factor is 3.18.

$$\text{New product variance} = (0.607/3.18)^2 = 0.036495 \qquad (2.23)$$

Table 2.5 Average Measurement Values (Average of Six Measurements) for Each Product and the Range of Products' Averages in Ohms after Retraining Operator 2

PRODUCT NO.	PRODUCT AVERAGE
P1	5.194
P2	4.905
P3	5.384
P4	5.253
P5	5.512
P6	5.242
P7	5.310
P8	5.125
P9	5.256
P10	5.270
Range	0.607

STEP 5: Calculate the total of three variances, namely gage repeatability, gage (test operator) reproducibility, and product variability, and the ratio of each variance to the total variance.

The total variance for the improved short-term gage capability analysis is given in Equation 2.24.

New total variance for this improved short-term gage capability evaluation = 0.000211 + 0.000041 + 0.036495 = 0.036747 (2.24)

Then we can obtain the ratio of each type of variance to the total variance as shown in Equations 2.25 through 2.28.

Improved gage repeatability variance
= (0.000211/0.036747) × 100 = 0.573% of total variance (2.25)

Improved gage reproducibility variance
= 0.111% of total variance (2.26)

Improved gage repeatability and gage reproducibility variance
= (0.573% + 0.111%) = 0.684% of total variance (2.27)

New product variance = 99.316% of total variance (2.28)

Both the gage repeatability and gage reproducibility variances improved a great deal. The gage repeatability variance decreased from 2.11% to 0.573%. The gage reproducibility variance decreased from 2.35% to 0.111%. Now let us see how much our measurement system's precision of the gage to total tolerance ratio improved.

STEP 6: Calculate the precision of the measurement system (gage plus operators) to the total tolerance ratio and decide if you can use this gage and work with these test operators on your production line or not.

The improved precision of the gage to total tolerance ratio becomes

Improved precision of gage to total tolerance ratio = 100
× [6 × square root (0.000211 + 0.000041)]/(5.5 − 4.5) = 9.51%
(2.29)

Retraining test operator 2 definitely did the trick. We reduced our gage capability for this measurement system to less than 10%. The next step is to find out if our measurement system for this parameter is capable in the long term. We can start using this gage and these three test operators on our production line. We can also start collecting data for the long-term gage capability analysis—discussed in Chapter 3—using the present gage, gage monitoring sample products, and our three qualified test operators.

Let us analyze another gage with 15 product samples and with one certified and experienced operator. This operator did repeated measurements five times on each product sample and the measurement data are presented in Table 2.6.

STEP 1: Calculate the ranges of measurements for every product and averages of ranges.

The gage capability evaluation for the data presented in Table 2.6 started with the measurement data sheet in the table. The range for each product's five measurements was calculated in the last column. The average of ranges for 15 products is 9.60 µV.

Table 2.6 Sample Short-Term Gage Capability Measurement Data Sheet for a Variable for Only Gage Repeatability Calculations

GAGE: Sensor Output Tester, Serial No. JHT12		DATE: October 2006			
PART NAME: Magnetic Sensor		PART NUMBER: S452601			
CRITICAL DIMENSION: Output at high frequency		TOLERANCE: 400 ± 80 µV			
OPERATOR: Jonathan					

PRODUCT ID	FIRST TRIAL	SECOND TRIAL	THIRD TRIAL	FOURTH TRIAL	FIFTH TRIAL	ABSOLUTE RANGE
S32	376	372	373	370	372	6
S41	340	333	341	325	356	31
S09	411	407	410	409	414	7
S122	405	402	405	401	403	4
S77	443	441	442	446	440	6
S65	462	467	463	461	468	7
S22	402	409	403	401	402	8
S94	382	384	380	379	381	5
S91	425	424	423	427	426	4
S87	409	403	425	431	436	33
S18	355	359	352	353	360	8
S71	433	431	436	435	431	5
S109	402	406	403	409	401	8
S66	388	391	385	387	390	6
S83	442	441	447	442	446	6
					Average of ranges	9.60

STEP 2: Calculate the gage repeatability variance.

Gage repeatability variance = [(Average of ranges for 15 products)/ (Factor for converting average range into gage repeatability variance, d_2^*)]^2 (2.30)

The factor for converting the average range into gage repeatability variance, d_2^*, is obtained from A. J. Duncan's *Quality Control and Industrial Statistics* (1974), p. 950, Table D3 or from Table 2.2 in this chapter. For this case, namely for 15 samples and 5 repeated measurement values in each range, the conversion factor is 2.34.

$$\text{Gage repeatability variance} = (9.60/2.34)^2 = 16.83 \quad (2.31)$$

STEP 3: We skip step 3 calculations for the reproducibility variance, as we have one test operator who is certified and experienced in using this gage.

STEP 4: Calculate the average of five measurements for each product, the range for these averages, and the product variance.

Next we need to determine the product variance in these 15 samples. The average values for each product, obtained from five measurements for that product, are shown in Table 2.6. The range of 15 products' averages is given in Table 2.7.

We are ready to calculate the product variance from the range of 15 products' averages by dividing it by a different factor, d_2^*, as shown in Equations 2.32 and 2.33.

Product variance = [(Range of products' averages)/(Factor for converting range of products' averages into product variance, d_2^*)]^2
        (2.32)

The factor for converting the range of products' averages into product variance, d_2^*, is again obtained from A. J. Duncan's *Quality Control and Industrial Statistics* (1974), p. 950, Table D3. For the present case, we use 1 measurement—the average of 5 measurements for each sample product—and the range for 15 sample products' averages, and the conversion factor is 3.55.

$$\text{Product variance} = (125.2/3.55)^2 = 1243.80 \quad (2.33)$$

STEP 5: Calculate the total of two variances, namely gage repeatability and product variability, and the ratio of each variance to the total variance.

Table 2.7 Average Values for Each Product (Average of Five Measurements) and the Range for 15 Products' Averages in µV

PRODUCT NO.	PRODUCT AVERAGE
S32	372.6
S41	339.0
S09	410.2
S122	403.2
S77	442.4
S65	464.2
S22	403.4
S94	381.2
S91	425.0
S87	420.8
S18	355.8
S71	433.2
S109	404.2
S66	388.2
S83	443.6
Range of averages	125.2

The total variance for the short-term gage capability analysis is the sum of two variances for this case, namely gage repeatability variance and product variance, as shown in Equations 2.34 and 2.35.

Total variance for short-term gage capability evaluation = Gage repeatability variance + Product variance   (2.34)

Total variance for this short-term gage capability evaluation
$$= 16.83 + 1243.80 = 1260.63 \qquad (2.35)$$

Then we can obtain the ratio of each type of variance to the total variance as shown in Equations 2.36 and 2.37.

Gage repeatability variance = $(16.83/1260.63) \times 100$
$$= 1.34\% \text{ of total variance} \qquad (2.36)$$

Product variance = 98.66% of total variance   (2.37)

As we can see from the preceding results, gage repeatability is only 1.34% of the total variance. Is this gage repeatability variance good enough for the precision of our measurement system? We analyze this question next.

STEP 6: Calculate the precision of the measurement system (one gage and a certified operator) to total tolerance ratio and decide if you can use this gage on your production line or not.

Precision of gage to total tolerance ratio = 100 × [6 × square root (Gage repeatability variance)]/(Total critical dimension tolerance)

$$(2.38)$$

If the precision of the gage to total tolerance ratio is less than 10%, then the gage that is being evaluated in measuring this critical parameter of our product is an acceptable one. The measurement system has a good enough precision to measure this critical parameter. If the precision of the gage to total tolerance ratio is between 10% and 30%, the gage that is being evaluated is doing a marginal job of measuring this critical parameter of our product. The measurement gage needs improvement and then repeated measurements have to be performed again to show a precision of gage to total tolerance ratio less than 10%. If the precision of gage to total tolerance ratio is greater than 30%, you might be wasting your time trying to improve this gage. It might be more advantageous to get a new gage.

Precision of gage to total tolerance ratio
= 100 × [6 × square root (16.83)]/(480-320) = 15.4% $\quad$ (2.39)

We see that the precision of the gage to total tolerance ratio given in Equation 2.39 for the measurement data in Table 2.6 is higher than 10%. We cannot use this gage to check this critical parameter for our product.

First we need to go over the measurement data to see if there are any anomalies. When we look at the measurement data in Table 2.6, we can easily see that test products S41 and S87 have repeatability problems. These two products have very high repeatability ranges and most likely they are unstable or malfunctioning. If we eliminate measurement data belonging to these two problematic products, then the remaining 13 products' measurement data seem more stable during testing and their repeatability ranges are comparable. Let us reanalyze the measurement data in Table 2.6 by eliminating measurement data for products S41 and S87.

The modified measurement data set is presented in Table 2.8.

STEP 1: Calculate ranges of measurements for every product and averages of ranges.

The gage capability evaluation for the data presented in the modified Table 2.8 started with the measurement data sheet in the table. The range for each product's five measurements was calculated in the last column. The average of ranges for 13 products is 6.15 µV.

STEP 2: Calculate the gage repeatability variance.

Gage repeatability variance = [(Average of ranges for 13 products)/ (Factor for converting average range into gage repeatability variance, d_2^*)]² (2.40)

The factor for converting average range into gage repeatability variance, d_2^*, is obtained from A. J. Duncan's *Quality Control and Industrial Statistics* (1974), p. 950, Table D3, or from Table 2.2 in this chapter.

Table 2.8 Modified Short-Term Gage Capability Measurement Data Sheet for a Variable for Only Gage Repeatability Calculations

GAGE: Sensor Output Tester, Serial No. JHT12   DATE: October 12, 2006
PART NAME: Magnetic Sensor   PART NUMBER: S452601
CRITICAL DIMENSION: Output at high   TOLERANCE: 400 ± 80 µV
  frequency
OPERATOR: Jonathan

PRODUCT ID	FIRST TRIAL	SECOND TRIAL	THIRD TRIAL	FOURTH TRIAL	FIFTH TRIAL	ABSOLUTE RANGE
S32	376	372	373	370	372	6
S09	411	407	410	409	414	7
S122	405	402	405	401	403	4
S77	443	441	442	446	440	6
S65	462	467	463	461	468	7
S22	402	409	403	401	402	8
S94	382	384	380	379	381	5
S91	425	424	423	427	426	4
S18	355	359	352	353	360	8
S71	433	431	436	435	431	5
S109	402	406	403	409	401	8
S66	388	391	385	387	390	6
S83	442	441	447	442	446	6
					Average of ranges	6.15

For this case, namely for 13 samples and 5 values in each range, the conversion factor is 2.34.

$$\text{Gage repeatability variance} = (6.15/2.34)^2 = 6.92 \qquad (2.41)$$

STEP 3: We skip step 3 calculation reproducibility calculations, as we have one test operator who is certified and experienced in using this gage.

STEP 4: Calculate the average of five measurements for each product, the range for these averages, and the product variance.

Next we need to determine the product variance in these 13 samples. The average values for each product, obtained from five measurements for that product, are shown in Table 2.9. The range of 13 products' averages is also given in Table 2.9.

We are ready to calculate the product variance from the range of 13 products' averages by dividing it by a different factor, d_2^*, as shown in Equations 2.42 and 2.43.

Product variance = [(Range of products' averages)/(Factor for converting range of products' averages into product variance, d_2^*)]^2

$$(2.42)$$

Table 2.9 Average Values for Each Product (Average of Five Measurements) and the Range of 13 Products' Averages in μV

PRODUCT NO.	PRODUCT AVERAGE
S32	372.6
S09	410.2
S122	403.2
S77	442.4
S65	464.2
S22	403.4
S94	381.2
S91	425.0
S18	355.8
S71	433.2
S109	404.2
S66	388.2
S83	443.6
Range of averages	108.4

The factor for converting the range of products' averages into product variance, d_2^*, is again obtained from A. J. Duncan's Quality Control and Industrial Statistics (1974), p. 950, Table D3. For the present case, we use 1 measurement—the average of 5 measurements for each sample product—and the range for 13 sample products' averages. The conversion factor here is 3.42.

$$\text{Product variance} = (108.4/3.42)^2 = 1004.63 \qquad (2.43)$$

STEP 5: Calculate the total of two variances, namely gage repeatability and product variability, and the ratio of each variance to the total variance.

The total variance for the short-term gage capability analysis is the sum of two variances for this case, namely gage repeatability variance and product variance, as shown in Equations 2.44 and 2.45.

Total variance for short-term gage capability evaluation
$$= \text{Gage repeatability variance} + \text{Product variance} \qquad (2.44)$$

Total variance for this short-term gage capability evaluation
$$= 6.92 + 1004.63 = 1011.55 \qquad (2.45)$$

Then we can obtain the ratio of each type of variance to the total variance as shown in Equations 2.46 and 2.47.

Gage repeatability variance is $(6.92/1011.55) \times 100$
$$= 0.68\% \text{ of total variance} \qquad (2.46)$$

$$\text{Product variance} = 99.32\% \text{ of total variance} \qquad (2.47)$$

As we can see from the preceding results, gage repeatability is reduced to 0.68% of the total variance. Is this gage repeatability variance good enough for the precision of our measurement system?

STEP 6: Calculate the precision of the measurement system (one gage and a certified operator) to the total tolerance ratio and decide if you can use this gage on your production line or not.

Precision of gage to total tolerance ratio = $100 \times [6 \times$ square root (Gage repeatability variance)]/(Total critical dimension tolerance)
$$(2.48)$$

After measurement data modifications, if the precision of the gage to total tolerance ratio is less than 10%, then the gage that is being evaluated in measuring this critical parameter of our product is an acceptable one.

$$\text{Precision of gage to total tolerance ratio}$$
$$= 100 \times [6 \times \text{square root } (6.92)]/(480 - 320) = 9.86\% \quad (2.49)$$

We see that precision of gage to total tolerance ratio given in Equation 2.49 for the modified measurement data in Table 2.7 is below 10%. By eliminating measurement data belonging to two anomalous product samples, we achieved an acceptable precision of the gage to total tolerance ratio.

The preceding examples show that reviewing the short-term gage capability data is very important. Anomalous data need to be removed from your calculations, if you have a good reason to believe those data are deviances. Otherwise, short-term gage capability measurements should be repeated.

3

LONG-TERM GAGE CAPABILITY EVALUATION FOR VARIABLES

After the short-term gage capability for measurement apparatus has been established with acceptable precision, you can start using it on your production line. However, this is not the end of the story for gage capability analysis. You have to also determine a gage's long-term stability and capability on your production line. Most long-term gage capability analyses are performed for a period of a month by recording measurement data each consecutive day. If you want to shorten the duration of long-term gage capability data collection, you can also take data in the morning once and in the afternoon once to cut the duration in half.

Choose at random a minimum of 10 monitoring products for long-term gage capability analysis. In the following example I use 10 monitoring products. The true values of these 10 monitoring products are shown in Table 3.1. The process specification used for this gage is 5.25 ± 0.25 Ohms. The true values of your monitoring products should cover the specification range and beyond, namely some reject products, too.

After measuring these 10 monitoring products daily using the same gage, a digital multimeter with serial number 90128, and the same qualified and experienced operator, namely Alice, for 25 days, I obtained the following data shown in Table 3.2.

We will treat the data in Table 3.2 using $\bar{X}$ and S charts for variables that are presented in Chapter 7. The averages and standard deviations—see Equations 7.1 through 7.4—of 10 monitor products ($P = 10$) for each measurement day ($M = 25$) are calculated and given

Table 3.1 True Values of 10 Monitoring Products (Sensors) Selected at Random for Long-Term Gage Capability Evaluation

MONITORING PRODUCT NO.	MONITORING PRODUCT TRUE VALUE IN OHMS
P1	4.234
P2	5.238
P3	5.537
P4	5.140
P5	5.353
P6	4.855
P7	4.678
P8	5.061
P9	4.988
P10	5.261

in Table 3.3. The average of 10 monitor products for measurement day 1 is shown as an example in Equation 3.1.

$$\text{Average of measurement day } 1 = \frac{1}{10} \times (4.234 + 5.201 + 5.509 + 5.205$$

$$+ 5.317 + 4.819 + 4.644 + 5.027 + 4.999 + 5211) = 5.017 \, \text{Ohms}$$

$$(3.1)$$

The standard deviation of 10 monitor products for measurement day 1 is shown as an example in Equation 3.2.

$$\text{Standard deviation of measurement day } 1 = \frac{1}{(10-1)^{(1/2)}} \Big[(4.234 - 5.017)^2$$

$$+ (5.201 - 5.017)^2 + (5.509 - 5.017)^2 + (5.205 - 5.017)^2$$

$$+ (5.317 - 5.017)^2 + (4.819 - 5.017)^2 + (4.644 - 5.017)^2$$

$$+ (5.027 - 5.017)^2 + (4.999 - 5.017)^2 + (5.211 - 5.017)^2 \Big]^{1/2}$$

$$= 0.370 \, \text{Ohms}$$

$$(3.2)$$

Then we have to calculate averages for 25 days of measurements, namely $\bar{\bar{X}}$ and $\bar{S}$, using Equations 7.2 and 7.4, respectively. These results are shown in the last row of Table 3.3. Next we have to calculate the standard deviation for 25 repeated measurements ($M = 25$) for

Table 3.2   Long-Term Gage Capability Measurement Data (in Ohms) Taken by Alice for Digital Multimeter with Serial No. 90128

DAY	MEASUREMENTS	P1	P2	P3	P4	P5	P6	P7	P8	P9	P10
9/5/2004	1	4.234	5.201	5.509	5.205	5.317	4.819	4.644	5.027	4.999	5.211
9/6/2004	2	4.241	5.225	5.556	5.202	5.302	4.835	4.638	5.035	4.988	5.201
9/7/2004	3	4.204	5.230	5.527	5.204	5.309	4.881	4.652	5.023	4.996	5.202
9/8/2004	4	4.276	5.230	5.533	5.204	5.313	4.829	4.607	5.019	4.987	5.229
9/9/2004	5	4.266	5.222	5.548	5.201	5.315	4.805	4.646	5.018	4.999	5.237
9/10/2004	6	4.293	5.211	5.523	5.213	5.307	4.804	4.616	5.022	4.982	5.212
9/11/2004	7	4.254	5.210	5.505	5.208	5.319	4.825	4.650	5.017	5.000	5.226
9/12/2004	8	4.240	5.243	5.521	5.206	5.303	4.813	4.622	5.032	4.990	5.201
9/13/2004	9	4.250	5.212	5.506	5.207	5.311	4.810	4.629	5.002	5.000	5.231
9/14/2004	10	4.275	5.212	5.508	5.254	5.202	4.811	4.610	5.075	4.995	5.292
9/15/2004	11	4.228	5.203	5.511	5.212	5.302	4.834	4.680	5.007	4.992	5.266
9/16/2004	12	4.284	5.245	5.506	5.202	5.311	4.866	4.626	5.030	4.998	5.270
9/17/2004	13	4.247	5.201	5.558	5.206	5.303	4.813	4.608	5.010	4.994	5.228
9/18/2004	14	4.282	5.243	5.503	5.225	5.312	4.827	4.601	5.017	4.984	5.228
9/19/2004	15	4.244	5.227	5.543	5.218	5.303	4.831	4.624	5.021	4.994	5.216
9/20/2004	16	4.220	5.234	5.553	5.210	5.308	4.824	4.647	5.032	4.994	5.208
9/21/2004	17	4.297	5.234	5.537	5.204	5.318	4.870	4.606	5.008	4.994	5.245
9/22/2004	18	4.238	5.205	5.549	5.207	5.302	4.860	4.646	5.021	4.987	5.233
9/23/2004	19	4.203	5.204	5.558	5.219	5.319	4.791	4.626	5.006	4.980	5.215
9/24/2004	20	4.201	5.243	5.511	5.213	5.316	4.804	4.620	5.006	4.992	5.225
9/25/2004	21	4.203	5.266	5.504	5.223	5.308	4.806	4.638	5.019	4.992	5.285
9/26/2004	22	4.256	5.204	5.553	5.200	5.312	4.848	4.647	5.008	4.986	5.231
9/27/2004	23	4.265	5.224	5.550	5.218	5.300	4.805	4.619	5.000	4.987	5.214
9/28/2004	24	4.269	5.247	5.537	5.212	5.302	4.825	4.612	5.025	4.997	5.202
9/29/2004	25	4.220	5.216	5.515	5.202	5.301	4.847	4.626	5.011	4.986	5.209

Table 3.3 Averages and Standard Deviations of 10 Monitor Products for Each Measurement Day

DAY	MEASUREMENTS	AVERAGE OF 10 MONITOR PRODUCTS, $\bar{X}$, USING EQUATION 7.1 IN OHMS	STANDARD DEVIATION OF 10 MONITOR PRODUCTS, S, USING EQUATION 7.3 IN OHMS
9/5/2004	1	5.017	0.370
9/6/2004	2	5.022	0.374
9/7/2004	3	5.023	0.376
9/8/2004	4	5.023	0.370
9/9/2004	5	5.026	0.371
9/10/2004	6	5.018	0.363
9/11/2004	7	5.021	0.365
9/12/2004	8	5.017	0.374
9/13/2004	9	5.016	0.369
9/14/2004	10	5.023	0.366
9/15/2004	11	5.023	0.370
9/16/2004	12	5.034	0.363
9/17/2004	13	5.017	0.379
9/18/2004	14	5.022	0.367
9/19/2004	15	5.022	0.375
9/20/2004	16	5.023	0.381
9/21/2004	17	5.031	0.366
9/22/2004	18	5.025	0.373
9/23/2004	19	5.012	0.390
9/24/2004	20	5.013	0.385
9/25/2004	21	5.024	0.387
9/26/2004	22	5.024	0.370
9/27/2004	23	5.018	0.374
9/28/2004	24	5.023	0.371
9/29/2004	25	5.013	0.374
AVERAGES FOR 25 DAYS OF MEASUREMENTS		$\bar{\bar{X}} = 5.021$ Ohms	$\bar{S} = 0.373$ Ohms

each of the 10 monitoring products. Standard deviations for repeated measurements are given in Table 3.4.

Averages in Table 3.3 for 10 monitoring products characterize the variability of repeated measurements for long-term gage capability analysis. We will use these averages in plotting the $\bar{X}$ chart along with the mean of these averages, namely $\bar{\bar{X}} = 5.021$ Ohms. However, standard deviations in Table 3.3 for 10 monitoring products characterize both variability of repeated measurements and

Table 3.4 Standard Deviations for 25 Repeated Measurements of 10 Monitoring Products, S_M, (in Ohms)

P1	P2	P3	P4	P5	P6	P7	P8	P9	P10
0.029	0.018	0.020	0.011	0.022	0.023	0.019	0.015	0.006	0.025

Average standard deviation for repeated measurements $\bar{S}_M = 0.019$ Ohms

variability between our monitoring products. We will use standard deviations given in Table 3.3 in plotting the S chart along with the mean of these standard deviations, namely $\bar{S} = 0.373$ Ohms. We will use special constants, A and B, in calculating the upper and lower control limits for $\bar{X}$ and S charts for our long-term gage capability study so that variability between monitoring products is eliminated. Equations 3.3 through 3.6 are used to calculate control limits for $\bar{X}$ and S charts utilized in the present long-term gage capability analysis.

$$\text{UCL}_{\bar{X}} = \bar{\bar{X}} + A \times \bar{S}_M \tag{3.3}$$

$$\text{LCL}_{\bar{X}} = \bar{\bar{X}} - A \times \bar{S}_M \tag{3.4}$$

$$\text{UCL}_S = \bar{S} + B \times \bar{S}_M \tag{3.5}$$

$$\text{LCL}_S = \bar{S} - B \times \bar{S}_M \tag{3.6}$$

Constants A and B are calculated from Equations 3.7 and 3.8 for M repeated measurements and for P monitoring products.

$$A = A_{3 \text{ for } P} \times \left(\frac{c_{4 \text{ for } P}}{c_{4 \text{ for } M}} \right) \tag{3.7}$$

$$B = (B_{4 \text{ for } P} - 1) \times \left(\frac{c_{4 \text{ for } P}}{c_{4 \text{ for } M}} \right) \tag{3.8}$$

where A_3, B_4, and c_4 are regular $\bar{X}$ and S variable control charts constants that can be found in A. J. Duncan's *Quality Control and Industrial Statistics* (1974), p. 968, Table M, or in Tables 7.1 and 7.2 of this book. For the present case the constants A and B are calculated in Equations 3.9 and 3.10 for $P = 10$ and $M = 25$.

$$A = 0.975 \times \left(\frac{0.9727}{0.9896} \right) = 0.9583 \tag{3.9}$$

$$B = (1.716 - 1) \times \left(\frac{0.9727}{0.9896} \right) = 0.7038 \tag{3.10}$$

The control limit constants A and B for long-term gage capability analysis using $\bar{X}$ and S charts are given in Tables 3.5 and 3.6 for different numbers of monitoring products, P, used and for different numbers of repeated measurements, M, performed.

Upper and lower control limits are given in Equations 3.11 through 3.14.

$$\mathrm{UCL}_{\bar{X}} = \bar{\bar{X}} + A \times \bar{S}_M = 5.021 + 0.9583 \times 0.019 = 5.0394 \tag{3.11}$$

$$\mathrm{LCL}_{\bar{X}} = \bar{\bar{X}} - A \times \bar{S}_M = 5.021 - 0.9583 \times 0.019 = 5.0032 \tag{3.12}$$

$$\mathrm{UCL}_S = \bar{S} + B \times \bar{S}_M = 0.373 + 0.7038 \times 0.019 = 0.3843 \tag{3.13}$$

$$\mathrm{LCL}_S = \bar{S} - B \times \bar{S}_M = 0.373 - 0.7038 \times 0.019 = 0.3577 \tag{3.14}$$

Table 3.5 Control Limit Constant A Used in Equations 3.3 and 3.4 for Different Numbers of Monitoring Products, P, and for Different Numbers of Repeated Measurements, M

	$P = 8$	$P = 9$	$P = 10$	$P = 11$	$P = 12$	$P = 13$	$P = 14$	$P = 15$
$M = 15$	1.0796	1.0183	0.9655	0.9205	0.8818	0.8475	0.8159	0.7890
$M = 16$	1.0783	1.0171	0.9643	0.9194	0.8807	0.8465	0.8149	0.7880
$M = 17$	1.0772	1.0161	0.9633	0.9184	0.8798	0.8456	0.8141	0.7872
$M = 18$	1.0762	1.0151	0.9624	0.9176	0.8790	0.8448	0.8134	0.7865
$M = 19$	1.0754	1.0143	0.9617	0.9168	0.8783	0.8441	0.8127	0.7859
$M = 20$	1.0746	1.0136	0.9610	0.9162	0.8777	0.8435	0.8121	0.7853
$M = 21$	1.0739	1.0129	0.9603	0.9155	0.8770	0.8429	0.8115	0.7848
$M = 22$	1.0732	1.0123	0.9597	0.9150	0.8765	0.8424	0.8110	0.7843
$M = 23$	1.0727	1.0118	0.9592	0.9145	0.8761	0.8420	0.8106	0.7839
$M = 24$	1.0721	1.0112	0.9587	0.9141	0.8756	0.8416	0.8102	0.7835
$M = 25$	1.0717	1.0108	0.9583	0.9137	0.8753	0.8412	0.8099	0.7832

Table 3.6 Control Limit Constant B Used in Equations 3.5 and 3.6 for Different Numbers of Monitoring Products, P, and for Different Numbers of Repeated Measurements, M

	$P = 8$	$P = 9$	$P = 10$	$P = 11$	$P = 12$	$P = 13$	$P = 14$	$P = 15$
$M = 15$	0.8006	0.7509	0.7090	0.6742	0.6429	0.6162	0.5932	0.5720
$M = 16$	0.7997	0.7500	0.7081	0.6734	0.6421	0.6154	0.5925	0.5713
$M = 17$	0.7989	0.7493	0.7074	0.6727	0.6415	0.6148	0.5919	0.5707
$M = 18$	0.7981	0.7486	0.7068	0.6721	0.6409	0.6142	0.5913	0.5702
$M = 19$	0.7975	0.7480	0.7062	0.6716	0.6404	0.6137	0.5909	0.5697
$M = 20$	0.7969	0.7474	0.7057	0.6711	0.6399	0.6133	0.5904	0.5693
$M = 21$	0.7963	0.7469	0.7052	0.6706	0.6395	0.6129	0.5900	0.5689
$M = 22$	0.7959	0.7464	0.7048	0.6702	0.6391	0.6125	0.5897	0.5686
$M = 23$	0.7955	0.7461	0.7044	0.6699	0.6387	0.6122	0.5894	0.5683
$M = 24$	0.7951	0.7457	0.7041	0.6695	0.6384	0.6119	0.5891	0.5680
$M = 25$	0.7947	0.7454	0.7038	0.6693	0.6382	0.6116	0.5888	0.5678

We have all the information to plot $\bar{X}$ and S charts for the long-term gage capability analysis. These charts are shown in Figures 3.1 and 3.2.

Before we can analyze results shown in Figures 3.1 and 3.2 for long-term gage capability, we have to determine if the $\bar{X}$ and S variable charts given earlier are in statistical control or not. Statistical control checks for $\bar{X}$ and S variable charts are the same as the ones presented in Chapter 6 for $\bar{X}$ and R variable control charts, namely

- Points outside control limits
- Eight points or more runs in a row on either side of the average
- Increasing or decreasing six intervals between points
- Cyclic nonrandom patterns
- Nonrandom patterns very close to the centerline (more than two-thirds of points on the control chart)
- Nonrandom patterns very close to the control limits (more than one-third of points on the control chart)

The $\bar{X}$ chart in Figure 3.1 seems to be in statistical control throughout all 25 measurement days. We do not have any situations that meet the aforementioned out-of-control conditions. However, in Figure 3.2, S is over the upper control limit for measurement days 19, 20, and 21. If we go through the data in Tables 3.2 and 3.3, we

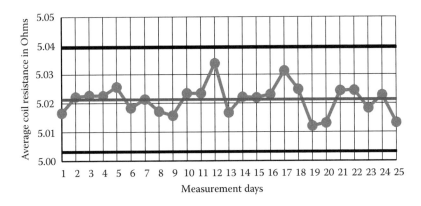

Figure 3.1   Average coil resistance for 10 monitoring products versus 25 measurement days.

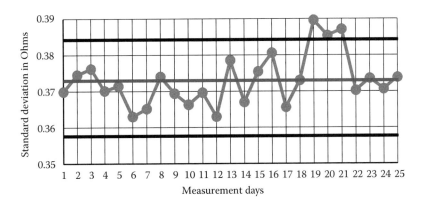

Figure 3.2   Standard deviation of coil resistance for 10 monitoring products versus 25 measurement days.

see that standard deviations of 10 monitoring products increased during days 19, 20, and 21. Our measurements were out of control during those three days. We have to determine special causes for out-of-control conditions during those three days. Once we find these special causes and take precautions so that they do not happen again, we can reanalyze the measurement data by disregarding data for those three days. Another option is to take more gage measurement data for few days to verify that special causes of variation are eliminated. In the present long-term gage capability analysis, let us use the stable data for 22 days and let us recalculate averages and upper and control limits for those 22 days. Here we assume that special causes for those out-of-control points have been determined and precautions are taken so that these special causes do not occur again. New values of $\bar{X}$, $\bar{S}$, $\bar{S}_M$, A, and B are

$$\bar{X} = 5.022 \text{ Ohms}, \bar{S} = 0.371 \text{ Ohms}, \bar{S}_M = 0.018 \text{ Ohms}$$

$$A = 0.9597 \text{ and } B = 0.7048 \tag{3.15}$$

New values for the upper and the lower control limits for $\bar{X}$ and S charts do not change much and they are given in Equations 3.16 and 3.17.

$$\text{UCL}_{\bar{X}} = 5.022 + 0.9597 \times 0.018 = 5.0394 \text{ Ohms and}$$
$$\text{LCL}_{\bar{X}} = 5.022 - 0.9597 \times 0.018 = 5.0044 \text{ Ohms} \tag{3.16}$$

$$\text{UCL}_S = 0.371 + 0.7048 \times 0.018 = 0.3838 \text{ and}$$
$$\text{LCL}_S = 0.371 - 0.7048 \times 0.018 = 0.3581 \tag{3.17}$$

We have all the information to replot $\bar{X}$ and S charts for the long-term gage capability analysis using 22 days worth of data. These charts are shown in Figures 3.3 and 3.4.

We see that the aforementioned $\bar{X}$ and S variable charts are in statistical control. We can now analyze the control charts shown in Figures 3.3 and 3.4 for long-term gage capability. The long-term gage repeatability standard deviation is obtained from the standard deviation that was obtained for 22 days of measurements in Equation 3.5 by using Equation 3.18.

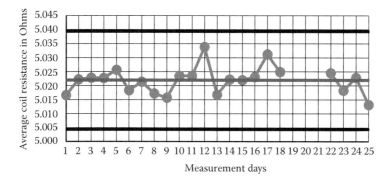

Figure 3.3 Average coil resistance for 10 monitoring products versus measurement days. (Days 19, 20, and 21 are deleted.)

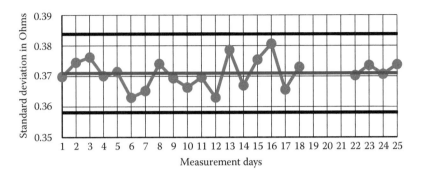

Figure 3.4 Standard deviation of coil resistance for 10 monitoring products versus measurement days. (Days 19, 20, and 21 are deleted.)

$$\sigma_{\text{long-term gage capability}} = \frac{\overline{S}}{c_4} \qquad (3.18)$$

The factor c_4 for converting 22 repeated measurements' standard deviations into long-term gage repeatability standard deviation is obtained from A. J. Duncan's *Quality Control and Industrial Statistics* (1974), p. 968, Table M, or from Table 7.2 in this book. For this case, namely for 22 measurements, the conversion factor is 0.9882.

$$\sigma_{\text{long-term gage capability}} = \frac{0.018}{0.9882} = 0.0184 \text{ Ohms} \qquad (3.19)$$

Assuming that long-term repeatability measurements we made for this gage are normally distributed (see Chapter 5) for a normal distribution description, then the long-term precision to tolerance ratio for this gage can be calculated. If this ratio is greater than 25%, then our gage is not a capable one in the long term.

Long-term precision to total tolerance ratio

$$= \frac{6 \times 100 \times \sigma_{\text{long-term gage capability}}}{\text{Upper Spec Limit} - \text{Lower Spec Limit}} \qquad (3.20)$$

$$\text{Long-term precision to total tolerance ratio} = \frac{6 \times 100 \times 0.0184}{5.5 - 5.0} = 22.1\% \qquad (3.21)$$

We see that our gage is a capable one in the long term since the long-term precision of our gage to tolerance ratio is less than 25%. I would like to emphasize again that special causes for those out-of-control points have been determined and precautions were taken so that these special causes do not occur again. Then we can enjoy the good result we obtained in Equation 3.21. We have to keep taking gage control measurement data daily. We have to plot the $\bar{X}$ and S data daily to make sure that our gage is in statistical control all the time for our production process measurements.

Let us perform another example of long-term gage capability evaluation by normalizing the repeatability data with respect to monitoring products' true values. In this example of long-term gage capability analysis, we will use only five monitoring products, namely $P = 5$. Because monitoring products are fewer than 8, we cannot use $\bar{X}$ and S variable control charts. Instead we will use $\bar{X}$ and R variable control charts; see Chapter 6. The gage in this example is used to measure a critical alignment dimension of a sensor. True values of five monitoring products that are used for this long-term gage capability evaluation are shown in Table 3.7. The specification used for this critical alignment dimension is 3250 ± 250 micrometers.

These monitoring products were measured twice a day for 15 days, namely $M = 30$. Measurement results are shown in Table 3.8.

Next we calculate the difference between each measurement and its true value. Let us call these differences Ds. Then we calculate the average of five differences for every measurement as shown in Equation 3.22.

$$\text{Averaged} = (D1 + D2 + D3 + D4 + D5)/5 \qquad (3.22)$$

Table 3.7 True Values of Five Monitoring Products (Sensors) in Micrometers for Long-Term Gage Capability Evaluation

MONITORING PRODUCT NO.	MONITORING PRODUCT TRUE VALUE IN MICROMETERS
P1	3222
P2	3676
P3	2980
P4	3367
P5	3111

Table 3.8 Alignment Measurements of Five Monitoring Products in Micrometers for Normalized Long-Term Gage Capability Evaluation for Gage No. LD-156

MEASUREMENT NO.	P1	P2	P3	P4	P5
1	3205	3670	3003	3353	3119
2	3246	3631	3005	3377	3111
3	3240	3667	2981	3347	3120
4	3216	3644	3001	3348	3117
5	3222	3637	2975	3389	3127
6	3228	3652	2995	3361	3126
7	3231	3645	2987	3359	3123
8	3201	3641	2984	3390	3110
9	3221	3641	2989	3369	3115
10	3208	3674	3006	3380	3137
11	3225	3640	2994	3347	3139
12	3245	3669	3014	3372	3136
13	3212	3667	2978	3345	3139
14	3209	3643	3010	3390	3131
15	3236	3650	3016	3351	3125
16	3233	3638	3019	3343	3125
17	3242	3657	3018	3386	3103
18	3209	3638	3001	3352	3134
19	3213	3636	2978	3376	3126
20	3227	3631	3016	3384	3118
21	3213	3670	3007	3361	3119
22	3249	3653	2976	3356	3106
23	3213	3675	2991	3359	3116
24	3213	3656	3017	3371	3101
25	3244	3650	2984	3365	3117
26	3205	3665	2997	3379	3107
27	3204	3678	3004	3365	3133
28	3246	3631	3003	3381	3136
29	3209	3678	3005	3388	3133
30	3201	3674	2982	3345	3116

We also determine the range of five differences, Ds for every measurement as shown in Equation 3.23.

$$\text{Ranged } D = \text{MAX} (D1, D2, D3, D4, D5) - \text{MIN} (D1, D2, D3, D4, D5) \qquad (3.23)$$

All differences, Ds, all averages for Ds, and all ranges for Ds for all 30 measurements are shown in Table 3.9. Also, $\bar{\bar{D}}$, S_D, and $\bar{R}$ are provided in the last two rows for control charting.

Table 3.9   All Differences, Ds, All Averages for Ds, and All Ranges for Ds for All 30 Measurements in Micrometers

MEASUREMENT NO.	D1	D2	D3	D4	D5	AVERAGE $\overline{D}$	RANGE D
1	−17	−6	23	−14	8	−1	40
2	24	−45	25	10	0	3	70
3	18	−9	1	−20	9	0	38
4	−6	−32	21	−19	6	−6	53
5	0	−39	−5	22	16	−1	62
6	6	−24	15	−6	15	1	39
7	9	−31	7	−8	12	−2	43
8	−21	−35	4	23	−1	−6	57
9	−1	−35	9	2	4	−4	43
10	−14	−2	26	13	26	10	39
11	3	−36	14	−20	28	−2	64
12	23	−7	34	5	25	16	41
13	−10	−9	−2	−22	28	−3	50
14	−13	−33	30	23	20	5	64
15	14	−26	36	−16	14	4	62
16	11	−38	39	−24	14	0	77
17	20	−19	38	19	−8	10	56
18	−13	−38	21	−15	23	−4	61
19	−9	−40	−2	9	15	−5	55
20	5	−45	36	17	7	4	82
21	−9	−6	27	−6	8	3	35
22	27	−23	−4	−11	−5	−3	50
23	−9	−1	11	−8	5	0	21
24	−9	−20	37	4	−10	1	57
25	22	−26	4	−2	6	1	48
26	−17	−11	17	12	−4	0	34
27	−18	2	24	−2	22	6	42
28	24	−45	23	14	25	8	70
29	−13	2	25	21	22	12	37
30	−21	−2	2	−22	5	−8	27
					$\overline{\overline{D}}$	1.23	$\overline{R} = 50.63$
					S_D	5.73	

Note: $\overline{\overline{D}}$, S_D, and $\overline{R}$ are also provided for control charting.

Upper and lower control limits for control charting the average of differences and ranges of differences are given in Equations 3.24 through 3.27.

$$\mathrm{UCL}\overline{D} = \overline{\overline{D}} + 3 \times S_D = 1.23 + 3 \times 5.73 = 18.4 \qquad (3.24)$$

$$\text{LCL}\overline{\overline{D}} = \overline{\overline{D}} - 3 \times S_D = 1.23 - 3 \times 5.73 = -16.0 \qquad (3.25)$$

$$\text{UCLR} = \mathcal{D}_4 \times \overline{R} = 2.115 \times 50.63 = 107.1 \qquad (3.26)$$

$$\text{LCLR} = \mathcal{D}_3 \times \overline{R} = 0.0 \times 50.63 = 0.0 \qquad (3.27)$$

The constants $\mathcal{D}_3$ and $\mathcal{D}_4$ in Equations 3.26 and 3.27 are obtained from A. J. Duncan's *Quality Control and Industrial Statistics* (1974), p. 968, Table M, or from Table 6.2 in this book. Do not confuse these constants D_3 and D_4 with differences, Ds, calculated above between each measurement and its true value in this example. Using these control limits and average differences and ranges from Table 3.9, we can plot our control charts for 30 measurements. These control charts are shown in Figures 3.5 and 3.6.

The following statistical control checks for $\overline{X}$ and R variable control charts (see Chapter 6) show that both of the control charts in Figures 3.5 and 3.6 are in control throughout 30 measurements.

- Points outside control limits
- Eight points or more runs in a row on either side of the average
- Increasing or decreasing six intervals between points
- Cyclic nonrandom patterns
- Nonrandom patterns very close to the centerline (more than two-thirds of points on the control chart)

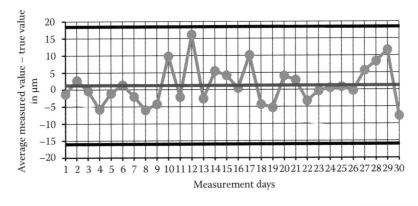

Figure 3.5 Average of five gage monitoring products measured value minus the true value in micrometers for 30 measurements.

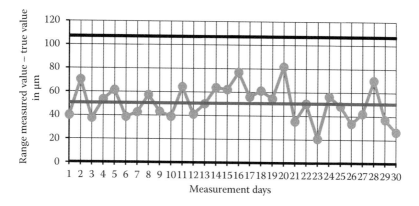

Figure 3.6   Range of five gage monitoring products measured value minus the true value in micrometers for 30 measurements.

- Nonrandom patterns very close to the control limits (more than one-third of points on the control chart)

As we have statistical control in both charts, we can go ahead and calculate our gage's long-term capability. For the gage capability calculations we need the average gage repeatability standard deviations for 30 measurements. This average standard deviation is obtained from the data calculated in Table 3.9 and is provided in Table 3.10.

The estimated long-term gage repeatability standard deviation is obtained from the 30 repeated measurements' standard deviation as follows:

$$\sigma_{\text{long-term repeatability}} = S_{M=30}/c_{4 \text{ for } M=30} \qquad (3.28)$$

The factor c_4 for converting 30 repeated measurements' standard deviations into long-term gage repeatability standard deviation is obtained from A. J. Duncan's *Quality Control and Industrial Statistics* (1974), p. 968, Table M, or from Table 7.2 in this book.

Table 3.10   Standard Deviations for 30 Repeated Measurements for 5 Monitoring Products in Micrometers

	D1	D2	D3	D4	D5
Standard deviation for 30 measurements in micrometers	15.591	15.818	13.846	15.621	10.944
Average standard deviation for 5 monitoring products and for $M=30$ measurements in micrometers $= 15.219$					

Note: The average standard deviation for 5 monitoring products and for 30 repeated measurements are also provided.

These sources provide the factor, c_4, up to $M = 25$. For larger values of M—repeated measurements—$c_4 \rightarrow 1.0$ or it can be approximated from Equation 3.29.

$$c_4 = 1 - \left(\frac{1}{4 \times M}\right) - \left(\frac{7}{32 \times M^2}\right) - \left(\frac{19}{128 \times M^3}\right) + O(M^{-4}) \quad (3.29)$$

For $M = 30$, $c_4 = 0.9914$. Then

$$\sigma_{\text{long-term repeatability}} = 15.219/0.9914 = 15.351 \quad (3.30)$$

Assuming that long-term repeatability measurements we made for this gage are normally distributed (see Chapter 5 for normal distribution description), then the long-term precision to tolerance ratio for this gage can be calculated. If this ratio is greater than 25%, then our gage is not a capable one in the long term.

Long-term precision to total tolerance ratio

$$= \frac{6 \times 100 \times \sigma_{\text{long-term gage capability}}}{\text{Upper Spec Limit} - \text{Lower Spec Limit}} \quad (3.31)$$

$$\text{Long-term precision to total tolerance ratio} = \frac{6 \times 100 \times 15.351}{3500 - 3000} = 18.4\%$$

$$(3.32)$$

We see that our gage is a capable one in the long term because the long-term precision of our gage to tolerance ratio is less than 25%. We still have to keep taking gage control measurement data daily. We have to keep plotting the average and the range of five gage monitoring products measured values minus the true values to make sure that our gage is in control all the time for our production process measurements.

4

Gage Capability for Attribute Data

Attribute type data have only two values: pass or fail, go or no-go, and so forth. An example of attribute data can be collected during final visual inspection of a product for cleanliness, chips, cracks, and metal marks. Clear-cut cases of acceptable, rejectable, marginally acceptable, and marginally rejectable products should be identified and photographed by responsible engineers. This is the most critical step for attribute data taking, namely developing precise definitions of conforming and nonconforming product conditions. You want to take human judgment out of attribute data as much as possible. When you are creating sample products as attribute standards for final visual inspection, you, your customer, and/or your subcontractor should all be in agreement. If you are trying to correlate with your customer and/or with your subcontractor, you all should have copies of photographs for these attribute standard products for final visual inspection with descriptive explanations so that there is no controversy when a product lot is rejected.

The second most critical step in gathering attribute data is to train your inspectors so that their repeatability and reproducibility of final inspection of your product result in very high percentages—very close to 100%.

The following steps will take you through attribute gage repeatability and reproducibility analysis for four final visual inspectors and compare their results to evaluation of these products by experienced engineers. Forty acceptable, rejectable, marginally acceptable, and marginally rejectable test products were chosen to identify cleanliness, chips, cracks, and metal marks. First these 40 test products were evaluated and classified, and their observations were denoted as "Pass" or "Fail" by experienced engineers. Then these 40 test products were inspected twice at random by each inspector. Each inspector successfully passed training earlier and had photographs of attribute standards for final visual inspection at his or her station. Inspectors did not know evaluation results of 40 test products from experienced engineers. The attribute data given in Table 4.1 were obtained.

Inspectors' Repeatability Calculations

Let us start with inspectors' repeatability calculations.

Two repeated inspections of inspector A agree with one another on 34 out of 40 parts or on (34/40) × 100 = 85% of test products.

Two repeated inspections of inspector B agree with one another on 37 out of 40 parts or on (37/40) × 100 = 92.5% of test products.

Two repeated inspections of inspector C agree with one another on 39 out of 40 parts or on (39/40) × 100 = 97.5% of test products.

Two repeated inspections of inspector D agree with one another on 39 out of 40 parts or on (39/40) × 100 = 97.5% of test products.

$$\text{Average repeatability for all four inspectors} \\ = (80\% + 92.5\% + 97.5\% + 97.5\%) = 93.1\% \qquad (4.1)$$

Our objective should be to bring all above repeatability percentages as close to 100% as possible for each inspector. Inspectors A and B definitely need retraining.

Four Inspectors' Reproducibility of Data

For reproducibility of attribute data, we look at agreement among all four inspectors and their two repeated assessments; that is, data in a row for the last eight columns in Table 4.1 should agree with each other. In Table 4.1, four inspectors agreed within their own assessments and with those of other inspectors 28 out of 40 times or in (28/40) × 100 = 70% of test products. 70% reproducibility is very low. We have to take immediate action to improve our final inspectors' understanding of "Pass" and "Fail" criteria.

Our objective should be to bring the reproducibility percentage as close to 100% as possible. From the repeatability and reproducibility analysis, we realize that inspectors A and B need retraining and then we have to repeat the attribute gage capability test for them.

Table 4.1 Final Inspection Attribute Data from 40 Test Products

TEST PRODUCT NO.	EXPERIENCED ENGINEERS DECISION	INSPECTOR A EVALUATION 1	INSPECTOR A EVALUATION 2	INSPECTOR B EVALUATION 1	INSPECTOR B EVALUATION 2	INSPECTOR C EVALUATION 1	INSPECTOR C EVALUATION 2	INSPECTOR D EVALUATION 1	INSPECTOR D EVALUATION 2
P1	Pass	Pass	Pass	Pass	Pass	Pass	Pass	Pass	Pass
P2	Pass	Pass	Pass	Pass	Pass	Pass	Pass	Pass	Pass
P3	Pass	Pass	Pass	Pass	Pass	Pass	Pass	Pass	Pass
P4	Pass	Pass	Pass	Pass	Pass	Pass	Pass	Pass	Pass
P5	Pass	Pass	Pass	Pass	Pass	Pass	Pass	Pass	Pass
P6	Fail	Fail	Pass	Fail	Fail	Fail	Fail	Fail	Fail
P7	Pass	Pass	Pass	Pass	Fail	Pass	Pass	Pass	Pass
P8	Fail	Fail	Pass	Fail	Fail	Fail	Fail	Fail	Fail
P9	Fail	Fail	Fail	Fail	Fail	Fail	Fail	Fail	Fail
P10	Fail	Fail	Fail	Fail	Fail	Pass	Fail	Fail	Fail
P11	Fail	Fail	Fail	Fail	Fail	Fail	Fail	Fail	Fail
P12	Pass	Pass	Pass	Pass	Pass	Pass	Pass	Pass	Pass
P13	Fail	Fail	Pass	Fail	Fail	Fail	Fail	Pass	Fail
P14	Fail	Fail	Fail	Fail	Fail	Fail	Fail	Fail	Fail
P15	Fail	Fail	Fail	Pass	Fail	Fail	Fail	Fail	Fail
P16	Pass	Pass	Pass	Pass	Pass	Pass	Pass	Pass	Pass
P17	Pass	Pass	Pass	Pass	Pass	Pass	Pass	Pass	Pass
P18	Pass	Pass	Pass	Pass	Pass	Pass	Pass	Pass	Pass
P19	Fail	Fail	Pass	Fail	Fail	Fail	Fail	Fail	Pass
P20	Fail	Fail	Fail	Fail	Fail	Fail	Fail	Fail	Fail

(Continued)

Table 4.1 (Continued) Final Inspection Attribute Data from 40 Test Products

TEST PRODUCT NO.	EXPERIENCED ENGINEERS DECISION	INSPECTOR A EVALUATION 1	INSPECTOR A EVALUATION 2	INSPECTOR B EVALUATION 1	INSPECTOR B EVALUATION 2	INSPECTOR C EVALUATION 1	INSPECTOR C EVALUATION 2	INSPECTOR D EVALUATION 1	INSPECTOR D EVALUATION 2
P21	Fail	Fail	Fail	Fail	Fail	Fail	Fail	Fail	Fail
P22	Fail	Fail	Fail	Fail	Fail	Pass	Pass	Fail	Fail
P23	Pass	Pass	Pass	Pass	Pass	Pass	Pass	Pass	Pass
P24	Fail	Fail	Fail	Fail	Pass	Pass	Pass	Fail	Fail
P25	Fail	Fail	Fail	Fail	Fail	Fail	Fail	Fail	Fail
P26	Fail	Fail	Fail	Fail	Fail	Fail	Fail	Fail	Fail
P27	Pass	Pass	Fail	Pass	Pass	Pass	Pass	Pass	Pass
P28	Pass	Pass	Pass	Pass	Pass	Pass	Pass	Pass	Pass
P29	Pass	Pass	Pass	Pass	Pass	Pass	Pass	Pass	Pass
P30	Pass	Fail	Fail	Fail	Fail	Pass	Pass	Pass	Pass
P31	Pass	Pass	Pass	Pass	Pass	Pass	Pass	Pass	Pass
P32	Pass	Pass	Pass	Pass	Pass	Pass	Pass	Pass	Pass
P33	Fail	Fail	Fail	Fail	Fail	Fail	Fail	Fail	Fail
P34	Fail	Fail	Fail	Fail	Fail	Fail	Fail	Fail	Fail
P35	Fail	Fail	Fail	Fail	Fail	Fail	Fail	Fail	Fail
P36	Fail	Pass	Fail	Fail	Fail	Fail	Fail	Fail	Fail
P37	Fail	Fail	Fail	Fail	Fail	Fail	Fail	Fail	Fail
P38	Pass	Fail	Fail	Fail	Fail	Fail	Fail	Fail	Pass
P39	Pass	Pass	Pass	Pass	Pass	Pass	Pass	Pass	Pass
P40	Pass	Pass	Pass	Pass	Pass	Pass	Pass	Pass	Pass

Each Inspector's Evaluation with Respect to Standard Assessment

Another aspect of this attribute analysis is to evaluate how well each inspector conducted the assessment with reference to him- or herself and with respect to experienced engineers' assessments. Inspectors might not understand "Pass" and "Fail" criteria for our product clearly.

Inspector A made the correct decision in both evaluations with respect to experienced engineers' assessments 32 out of 40 times or in (32/40) × 100 = 80% of test products.

Inspector B made the correct decision in both evaluations with respect to experienced engineers' assessments 35 out of 40 times or in (35/40) × 100 = 87.5% of test products.

Inspector C made the correct decision in both evaluations with respect to experienced engineers' assessments 36 out of 40 times or in (36/40) × 100 = 90% of test products.

Inspector D made the correct decision in both evaluations with respect to experienced engineers' assessments 37 out of 40 times or in (37/40) × 100 = 92.5% of test products.

These results show that our final inspection criteria are not fully understood by our four inspectors.

All Inspectors' Evaluations with Respect to Standard Assessment

All inspectors agree within themselves and with each other and with experienced engineers' decisions 27 out of 40 times, namely in (27/40) × 100 = 67.5% of test products. This three way agreement is true for test product numbers P1, P2, P3, P4, P5, P9, P11, P12, P14, P16, P17, P18, P20, P21, P23, P25, P26, P28, P29, P31, P32, P33, P34, P35, P37, P39, and P40. This result shows that our final inspection criteria are not fully understood by our inspectors. We can see misunderstandings with regard to test products P24, P30, and P38. Some specifications for inspection have to be defined more clearly and our inspectors have to be retrained to improve the effectiveness of our final visual inspection process. Our objective should be to bring all of the aforementioned percentages as close to 100% as possible.

We should also repeat the preceding attribute gage repeatability, reproducibility, and standards assessment test periodically, namely once a week or at most every two weeks, to ensure that our inspectors are not deviating from our standard criteria and are performing close to 100%.

Let us do another gage capability analysis for attribute data for an automated laser measurement system. Our products are introduced to a laser beam by automated fixtures. As a result of the laser detection system the products are sorted into three different bins. One bin is for "above limit" products, namely above the upper specification limit. The second bin is for "good" or within specification products. The third bin is for "below limit" products, namely below the lower specification limit. We have three such automated product sorting gages. We check them every day for their capability. During one such check using 20 gage monitoring products, we obtained the attribute data given in Table 4.2.

Three Gages' Repeatability Calculations

Let us start with the gages' repeatability calculations.

For automated gage 1, three repeated sortings at random for 20 gage monitor products agree with one another 18 out of 20 times or in $(18/20) \times 100 = 90\%$ of monitor products.

For automated gage 2, three repeated sortings at random for 20 gage monitor products agree with one another 20 out of 20 times or in $(20/20) \times 100 = 100\%$ of monitor products.

For automated gage 3, three repeated sortings at random for 20 gage monitor products agree with one another 20 out of 20 times or in $(20/20) \times 100 = 100\%$ of monitor products.

$$\text{Average repeatability for all three automated gages}$$
$$= (90\% + 100\% + 100\%) = 93.3\% \qquad (4.2)$$

Our objective should be to bring all of the repeatability percentages in the preceding text as close to 100% as possible for every automated gage. Gage 1 has an issue in sorting two "good" gage monitor products as "above limit." Gage 1 should not be used in production. Gage 1 should be reported to maintenance with a report that provides results of the aforementioned repeatability analysis.

Three Gages' Reproducibility of Data

For reproducibility of attribute data, we look at agreement among all three gages and their three repeated assessments; that is, data in a row

Table 4.2   Attribute Data for Three Automated Product Sorting Systems Using 20 Gage Monitoring Standards

	MONITOR STANDARD	GAGE 1 TRIAL 1	GAGE 1 TRIAL 2	GAGE 1 TRIAL 3	GAGE 2 TRIAL 1	GAGE 2 TRIAL 2	GAGE 2 TRIAL 3	GAGE 3 TRIAL 1	GAGE 3 TRIAL 2	GAGE 3 TRIAL 3
P1	*Good*	Good	Good	Good	Good	Good	Good	Good	Good	Good
P2	*Above limit*	Above limit	Above limit	Above limit	Above limit	Above limit	Above limit	Above limit	Above limit	Above limit
P3	*Good*	Good	Good	Good	Good	Good	Good	Good	Good	Good
P4	*Good*	Good	Good	Good	Good	Good	Good	Good	Good	Good
P5	*Below limit*	Below limit	Below limit	Below limit	Below limit	Below limit	Below limit	Good	Good	Good
P6	*Good*	Good	Good	Good	Good	Good	Good	Good	Good	Good
P7	*Above limit*	Above limit	Above limit	Above limit	Above limit	Above limit	Above limit	Above limit	Above limit	Above limit
P8	*Below limit*	Below limit	Below limit	Below limit	Below limit	Below limit	Below limit	Below limit	Below limit	Below limit
P9	*Below limit*	Below limit	Below limit	Below limit	Below limit	Below limit	Below limit	Good	Good	Good
P10	*Below limit*	Below limit	Below limit	Below limit	Below limit	Below limit	Below limit	Good	Good	Good
P11	*Above limit*	Above limit	Above limit	Above limit	Above limit	Above limit	Above limit	Above limit	Above limit	Above limit
P12	*Above limit*	Above limit	Above limit	Above limit	Above limit	Above limit	Above limit	Above limit	Above limit	Above limit
P13	*Good*	Good	Good	Good	Good	Good	Good	Good	Good	Good
P14	*Good*	Good	Good	Good	Good	Good	Good	Good	Good	Good
P15	*Above limit*	Above limit	Above limit	Above limit	Above limit	Above limit	Above limit	Above limit	Above limit	Above limit
P16	*Good*	Good	Good	Good	Good	Good	Good	Good	Good	Good
P17	*Good*	Good	Good	Good	Good	Good	Good	Good	Good	Good
P18	*Good*	Good	Good	Good	Good	Good	Good	Good	Good	Good
P19	*Below limit*	Below limit	Below limit	Below limit	Below limit	Below limit	Below limit	Good	Good	Good
P20	*Good*	Good	Good	Above limit	Good	Good	Good	Good	Good	Good

for the last nine columns in Table 4.2 should agree with each other. In Table 4.2, three gages agreed within themselves and with each other 14 out of 20 times or in (14/20) × 100 = 70% of gage monitoring standards. 70% reproducibility is very low. We have issues with gages 1 and 3. If we can correct the repeatability issue of gage 1 at the high end of "good" products and if we can recalibrate gage 3 for the low end of "good" products, we might be able to bring all three gages to have 100% repeatability and reproducibility.

Each Gage's Evaluation with Respect to Gage Monitoring Standards

Another aspect of this attribute analysis is to evaluate how well each gage performed within itself and with respect to the gage monitoring standards.

Gage 1 sorted correctly in all three evaluations with respect to gage monitoring standards' results 18 out of 20 times or in (18/20) × 100 = 90% of monitoring standards.

Gage 2 sorted correctly in all three evaluations with respect to gage monitoring standards' results 20 out of 20 times or in (20/20) × 100 = 100% of monitoring standards.

Gage 3 sorted correctly in all three evaluations with respect to gage monitoring standards' results 16 out of 20 times or (16/20) × 100 = 80%.

Gage 3 sorted four "below limit" monitoring standards as "good." We have to find the cause(s) for this misinterpretation in gage 3 and correct for it.

All Gages' Evaluations with Respect to Gage Monitoring Standards

All three gages agree within themselves and with each other and with monitor standards' results 14 out of 20 times, namely in (14/20) × 100 = 70%. All three gages agree within themselves and with each other and with monitor standards for P1, P2, P3, P4, P6, P7, P8, P11, P12, P13, P14, P15, P17, and P18. These results show that only gage 2 can be used in production as a result of this attribute gage capability analysis. Gages 1 and 3 have to be maintained and retested similar to the attribute test presented here before introducing them into volume production. Our objective should be to have all three gages operating at 100% for repeatability, reproducibility, and for correlating to gage monitor standards.

We should also repeat the aforementioned attribute gage repeatability, reproducibility, and standards assessment test periodically, namely once a week or at most every two weeks, to ensure that our gages are sorting 100% correctly.

5

NORMAL DISTRIBUTION

In probability theory and statistics there are many distribution functions that fit well to describe observations of a physical phenomenon. Distribution functions come with different averages, dispersions (spreads), and shapes. Binomial, Poisson, exponential, hypergeometric, chi-squared, student's t, and so forth distributions and their applications can be found in many probability and statistics books. The most applicable and important frequency distribution that fits observed and variable data in high-volume manufacturing and that helps us generate process control charts by sampling our product is the normal distribution. There are two main reasons why we focus on the normal distribution for process control charts.

1. During high-volume process control, distribution of product sampling averages can be approximated by a normal distribution.
2. The mean of the product sampling averages equals the total product's mean.

This chapter focuses on characteristics of this very important distribution function for our high-volume process control by providing simple examples.

The normal distribution function is bell shaped and it is perfectly symmetrical about the product mean. The normal distribution function is given by Equation 5.1, where X_{AVE} is the average and σ is the standard deviation of all measurements (whole population) for the product variable.

$$\text{Normal distribution for } P(X) = \left[\frac{1}{\sigma\sqrt{2\pi}}\right]\exp\left[-\frac{(X-X_{AVE})^2}{2\sigma^2}\right] \quad (5.1)$$

The normal distribution for $X_{AVE} = 0$ and $\sigma = 1$ is shown in Figure 5.1.

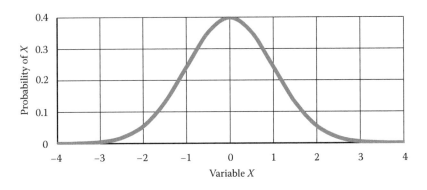

Figure 5.1 Normal distribution with $X_{AVE} = 0$ and $\sigma = 1$.

All products for this variable fall into an area under the normal distribution curve as illustrated in Figure 5.2. If we integrate the normal distribution curve in Figure 5.1 from $X \to -\infty$ to $X \to = +\infty$ we cover 100% of the product for this variable; that is, the probability of our product falling within $X \to -\infty$ to $X \to +\infty$ is 100%. Since the normal distribution is symmetrical about X_{AVE}, 50% of the product falls into the area from $X \to -\infty$ to $X = 0$ and 50% of the product falls into the area from $X = 0$ to $X \to +\infty$. These areas under the normal distribution function—called cumulative probabilities—can be found in statistics books in the form of tables. A good reference is A. J. Duncan's *Quality Control and Industrial Statistics* (1974),

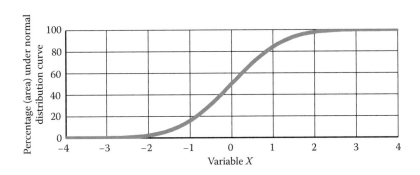

Figure 5.2 Area under the normal distribution function (cumulative probabilities) from $X \to -\infty$ to $X \to +\infty$ as a function of variable X with $X_{AVE} = 0$ and $\sigma = 1$.

p. 945, Table A2. Cumulative probabilities of the normal distribution are mostly given as a function of a nondimensional variable Z where $Z = (X - X_{AVE})/\sigma$. You can also obtain areas under the normal probability distribution by using the NORMDIST statistical formula in MS Excel®.

For example, using MS Excel, the area from $X \rightarrow -\infty$ to $X = 0$ with $X_{AVE} = 0$ and $\sigma = 1$ is NORMDIST(0,0,1, TRUE) = 0.5, which is the cumulative probability or 50% of the area under the normal probability distribution.

For any value of X_{AVE} and standard deviation, σ, the total area under the normal distribution function is one or it covers 100% of the product under investigation. Areas on each side of the average are equal. 68.26% of the product lies within $X = \pm 1\sigma$ from the average. 95.44% of the product lies within $X = \pm 2\sigma$ from the average. 99.74% of the product lies within $X = \pm 3\sigma$ from the average.

If the product's process for a variable is capable to $X = \pm 3\sigma$ from the average, then 2700 products out of a million that we produce will be out of specification. We can obtain this out-of-specification value by finding the area under the normal distribution curve from $X \rightarrow -\infty$ to $X = -3$ as shown in Equation 5.2 for $X_{AVE} = 0$ and $\sigma = 1$.

$$\text{Area under normal distribution curve from } X \rightarrow -\infty \text{ to}$$
$$X = -3 = \text{NORMDIST}(-3, 0, 1 \text{ TRUE}) = 0.001350 \qquad (5.2)$$

Since the normal distribution curve is symmetrical, the area under the normal distribution curve from $X = +3$ to $X \rightarrow +\infty$ is also 0.001350. The total number of products out of a million that fall outside the $X = \pm 3\sigma$ process capability is

$$\text{Total number of products out of } X = \pm 3\sigma \text{ process capability}$$
$$= 1{,}000{,}000 \times 2 \times 0.001350 = 2700 \qquad (5.3)$$

However, if we can achieve a capability of $X = \pm 4.5\sigma$ for this variable, then only 6.8 products out of a million that we produce will be out of specification. $X = \pm 4.5\sigma$ is a variable's process capability target for 6σ quality programs. Note that the remaining $\pm 1.5\sigma$ from the 6σ quality target is allocated to shifts in the product's average value.

Let us investigate the diameter of a pin that we produce for our customer in high-volume production. The diameter of the pin is a critical dimension specified as 1.000 ± 0.005 inches. We take five samples at random per shift to control our production process. We want to be a vendor qualified as 6σ ship-to-stock for this product so our target process capability for this critical dimension is X_{AVE} ± 4.5σ. We do not need the ±1.5σ variation for our product's X_{AVE}, as our X_{AVE} is very steady and it is 1.000 inch. Our pin diameter average is

$$X_{AVE} = 1.000 \text{ inch} \tag{5.4}$$

And our target pin diameter standard deviation is

$$\text{Process standard deviation } \sigma = 0.005/4.5 = 0.0011 \text{ inch} \tag{5.5}$$

The normal distribution of all populations of pin diameters using Equations 5.4 and 5.5 is shown in Figure 5.3. However, we cannot measure every pin diameter in high-volume production. Therefore we have to take samples at random and see if our high-volume production process is in control or not. Also, the spread of pin diameters for our production population should be less than 0.0011 inch. The standard deviation of sampling averages—for a sample size of 5—can be obtained from the desired population standard deviation as follows.

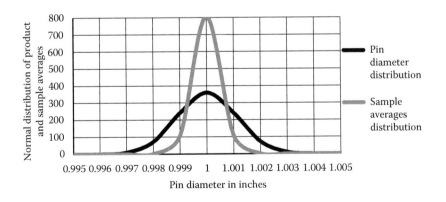

Figure 5.3 Distributions of all pin diameters and production control sample averages (for a sample size of 5).

Since the population of pin diameters is normally distributed, sampling averages are also normally distributed (see Figure 5.3), or

$$\text{Standard deviation of sampling averages}$$
$$= \text{Standard deviation of population}/\sqrt{N} \qquad (5.6)$$

for a sampling size of N. For our case, the sampling size is 5. Then

$$\text{Standard deviation of sampling averages} = \text{Standard deviation}$$
$$\text{of population}/\sqrt{5} = 0.0011 / \sqrt{5} = 0.0005 \text{ inches} \qquad (5.7)$$

We emphasized at the beginning of this chapter that when sampling from normal populations, the mean of the product sampling averages equals the total product's population mean. So using results from Equations 5.4, 5.5, and 5.7, we obtain the population and sampling distributions shown in Figure 5.3.

Sampling the distribution of averages for a variable in high-volume production will provide us with a very powerful tool to control our processes. Process control charts by sampling such as $\bar{X}$ and R—sample average and sample range charts—and $\bar{X}$ and S—sample average and sample standard deviation charts—will be discussed in detail in Chapters 6 and 7.

Let us further analyze the diameter of the pin for our product with X_{AVE} = 1.000 inches and standard deviation = 0.0011 inches. Figure 5.4 shows areas under the normal distribution for the pin diameter: $-X_{AVE} \pm 1\sigma$ covers 68.3% of the total area, $X_{AVE} \pm 2\sigma$ covers 95.5% of the total area, $X_{AVE} \pm 3\sigma$ covers 99.7% of the total area, and $X_{AVE} \pm 4\sigma$ covers 99.994% of the total area.

Out of a million products, 2700 will fall outside of $X_{AVE} \pm 3\sigma$ limits; namely, 1350 pins will have diameters less than 0.9967 inches and 1350 pins will have diameters greater than 1.0033 inches. 63 products out of a million will fall outside of $X_{AVE} \pm 4\sigma$ limits, namely 32 pins will have diameters less than 0.9956 inch and 32 pins will have diameters greater than 1.0044 inches. Only seven products out of a million will fall outside the $X_{AVE} \pm 4.5\sigma$ limits, which are our high-volume production targets, and three pins will have diameters less than 0.99505 inches and three pins will have diameters greater than 1.00495 inches.

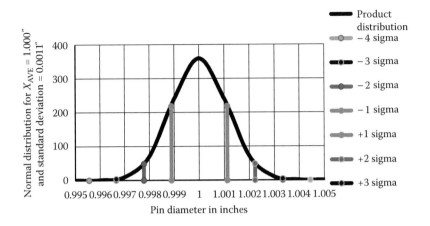

Figure 5.4 Normal distribution of pin diameters and areas (cumulative probabilities) under $X_{AVE} \pm 1\sigma = 68.3\%$, $X_{AVE} \pm 2\sigma = 95.5\%$, $X_{AVE} \pm 3\sigma = 99.7\%$, and $X_{AVE} \pm 4\sigma = 99.994\%$.

Areas (cumulative probabilities) under $X_{AVE} \pm 1\sigma$, $X_{AVE} \pm 2\sigma$, and $X_{AVE} \pm 3\sigma$ are clearly shown in Figures 5.5 through 5.7.

Let us do several examples with the above pin diameters, namely using $X_{AVE} = 1.000$ inch and population standard deviation = 0.0011 inch.

Example 1

One of our customers wants to know how many out of 200,000 pins that they will receive will have diameters less than 0.997 inch.

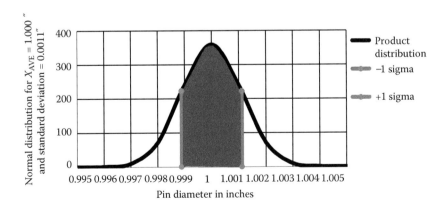

Figure 5.5 Normal distribution of all pin diameters and highlighted area under $X_{AVE} \pm 1\sigma = 68.3\%$.

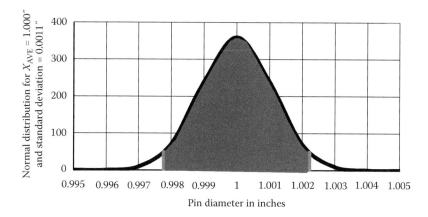

Figure 5.6 Normal distribution of all pin diameters and highlighted area under $X_{AVE} \pm 2\sigma =$ 95.5%.

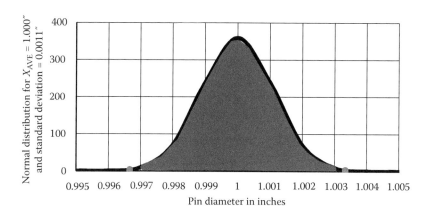

Figure 5.7 Normal distribution of all pin diameters and highlighted area under $X_{AVE} \pm 3\sigma =$ 99.7%.

By using MS Excel's NORMDIST function, we can get the answer easily by finding the area under the normal distribution from $X \to -\infty$ to $X = 0.997$ inch.

$$\text{NORMDIST}(0.997, 1, 0.0011, \text{TRUE})$$
$$= 0.003467 \text{ or } 0.35\% \tag{5.8}$$

or the number of pins with diameters below 0.997 inches is

$$200{,}000 \times 0.003467 = 693 \tag{5.9}$$

which is also shown in Figure 5.8.

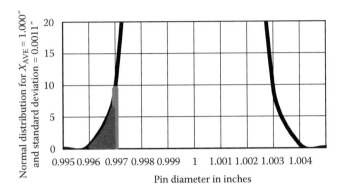

Figure 5.8   Number of pins with a diameter less than 0.997 inch is 693. The normal distribution area below 0.997 inch is highlighted.

Example 2

Another customer wants to know what percentage of a large quantity of pins that they will receive will have diameters between 1.002 and 1.003 inches.

By using MS Excel's NORMDIST function, we can get the answer easily by finding the area under the normal distribution from $X \to -\infty$ to 1.003 inches and by subtracting the area from $X \to -\infty$ to 1.002 inches.

$$\text{NORMDIST}(1.003,1,0.0011,\text{TRUE}) -$$
$$\text{NORMDIST}(1.002,1,0.0011,\text{TRUE})$$
$$= 0.032463 \text{ or } 3.25\% \qquad (5.10)$$

which is also shown in Figure 5.9.

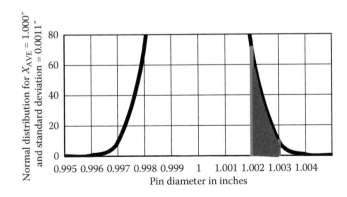

Figure 5.9   Normal distribution area of pins the customer will receive with diameters between 1.002 and 1.003 inches.

Example 3

We produce sensors with an average output of 240 mV and a population standard deviation of 10 mV. Our sensor's population output behaves like a normal distribution. A customer wants 10,000 sensors per week from us. The customer's specification for sensor output is a 230 mV minimum. How many sensors should we manufacture per week and then sort them 100% to meet this customer's delivery requirement?

We first find the percentage of products that fall between $X \rightarrow -\infty$ and $X = 230$ mV using MS Excel as follows and also as shown in Figure 5.10.

$$100 \times \text{NORMDIST}(230,240,10,\text{TRUE}) = 15.9\% \quad (5.11)$$

The number of sensors that need to be built every week to meet this customer's 10,000 sensors per week requirement is

$$\text{Weekly build quantity} = 10,000/[1 - (15.9/100)] = 11,886 \quad (5.12)$$

The highlighted area in Figure 5.10, which is 100% − 15.9% = 84.1%, represents sensors with outputs greater than 230 mV.

Example 4

We produce sensors with an average output of 240 mV and a population standard deviation of 10 mV similar to the one in

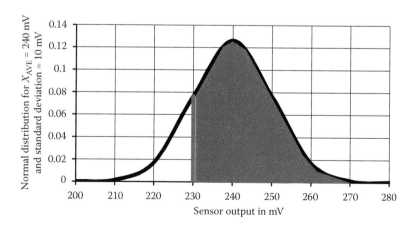

Figure 5.10 Normal distribution area of sensors with an output greater than 230 mV.

Example 3. I would like to take out 25 sensors at random from a large production population. What is the probability that the average output of these 25 sensors will lie between 239 mV and 244 mV?

First we have to determine the standard deviation for 25 sampling averages. Since our sensor output behaves like a normal distribution, then

Standard deviation of 25 sampling averages

$$= \text{Standard deviation of population}/\sqrt{25} = 10/\sqrt{25} = 2 \text{ mV} \quad (5.13)$$

Also, the 25 sampling distribution average is the same as the population average, which is

$$\text{Sampling distribution average} = 240 \text{ mV} \quad (5.14)$$

Now we can draw the sampling averages normal distribution by using Equations 5.13 and 5.14 and show the area between 239 mV and 244 mV as highlighted in Figure 5.11. This area represents the probability of getting the average of 25 randomly selected samples within 239 mV and 244 mV.

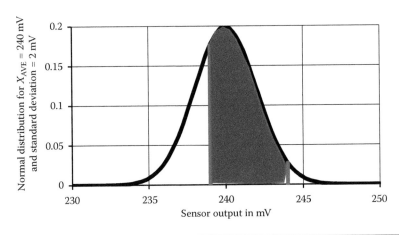

Figure 5.11 Sampling normal distribution for 25 samples and the highlighted area between an output of 234 mV and 244 mV.

The area between 239 mV output and 244 mV output can be determined by using the NORMDIST function in MS Excel as follows:

$$\text{NORMDIST}(244,240,2,\text{TRUE}) - $$
$$\text{NORMDIST}(239,240,2,\text{TRUE}) = 0.669 \qquad (5.15)$$

Then the probability that the average of our randomly chosen 25 samples will fall between 239 mV and 244 mV is

$$\text{Probability} = 100 \times 0.669 = 66.9\% \qquad (5.16)$$

6

PROCESS CONTROL FOR VARIABLES: $\bar{X}$ AND R CHARTS

In high-volume production it is not feasible and cost effective to measure each part to see if it complies with a given specification. Control charts for variables are simple and effective ways to achieve statistical control of your processes and to show your customers that parts you deliver to them are in a given specification. When a high-volume production is in statistical control, you will be able to predict how many parts will be out of your customer's specification. You will be able to achieve ship-to-stock status with your customers. Most high-volume production companies target 3.4 defective parts out of a million parts produced—six sigma quality target—in consistent and predictable production processes.

With accurate and consistent control charting, you can immediately observe when your production processes go out of statistical control. You can stop production and correct the cause(s) of variation. At the initial get-go, your production processes might not achieve the six sigma quality target. However, control charts that are in statistical control will help you to see how your process modifications are helping you to achieve your quality goals by reducing variability. You can perform design-of-experiments to normal causes of variation in your production processes and determine the optimum window of operation to achieve your six sigma quality target.

Variable control charts will provide information about the average and spread of a measurable variable in our process. First we have to have a capable measurement system (see Chapters 1 through 3) for the variable in question. Then we should start taking samples at random—sample sizes should be fixed between 2 and 10—at certain intervals that represent our processes. The interval for taking samples should be determined such that we can catch variations in our processes. A critical grinding process might require measurement samples taken every hour and/or before and after a new raw material lot. A wafer fabrication process might require measurement samples taken after every deposition lot or before and after changing a photoresist batch. Our goal is to detect variability in our manufacturing processes over time.

Initially we should collect at least 25 sets of data before calculating statistical parameters such as control limits for the variable control chart. For a sample size of 5, the initial data set for a variable control chart is shown in Table 6.1.

The average of five samples for Sample Group 1 is obtained as follows:

$$\bar{X}_1 = (24.85 + 24.92 + 25.10 + 24.56 + 24.60) / 5 = 24.806 \text{ Ohms} \quad (6.1)$$

The range of five samples for Sample Group 1 is obtained as follows:

$$R_1 = X_{\text{High for Sample Group 1}} - X_{\text{Low for Sample Group 1}}$$
$$= 25.10 - 24.56 = 0.54 \text{ Ohms} \quad (6.2)$$

Then we calculate the process average, $\bar{\bar{X}}$, and the process range, $\bar{R}$, for all sample groups where j is the number of sample groups. For the example given in Table 6.1, $j = 25$.

$$\bar{\bar{X}} = \frac{(\bar{X}_1 + \bar{X}_2 + \bar{X}_3 + \cdots + \bar{X}_{j-1} + \bar{X}_j)}{j} \quad (6.3)$$

$$\bar{R} = \frac{(R_1 + R_2 + R_3 + \cdots + R_{j-1} + R_j)}{j} \quad (6.4)$$

For the process data in Table 6.1, $\bar{\bar{X}} = 24.941$ Ohms and $\bar{R} = 0.546$ Ohms. Next we have to determine control limits for the average chart and for the range chart. Calculations for control limits in $\bar{X}$ and R variable charts use constants and these constants vary according to the sample size. The upper control limits (UCLs) and lower control limits (LCLs) in $\bar{X}$ and R variable charts are defined as follows:

$$\text{UCL}_{\bar{X}} = \bar{\bar{X}} + A_2 \times \bar{R} \quad \text{and} \quad \text{LCL}_{\bar{X}} = \bar{\bar{X}} - A_2 \times \bar{R} \quad (6.5)$$

Table 6.1 Initial $\bar{X}$ and R Control Chart Data for Thermal Sensor Resistance

PRODUCT NAME AND PART NO.: THERMAL SENSOR 6773-11	PROCESS: FINAL ASSEMBLY			VARIABLE: SENSOR RESISTANCE IN OHMS	SPECIFICATION LIMITS = 25±1 OHM				SAMPLE SIZE = 5 SAMPLE FREQUENCY = ONCE A SHIFT (1 HOUR AFTER THE START OF A SHIFT)	
Sample Group No.	1	2	3	4	...	23	24	25
Sample Time	09:00	17:00	01:00	09:00	...	17:00	01:00	09:00
Sample Date	6/2/2009	6/2/2009	6/3/2009	6/3/2009	...	6/9/2009	6/10/2009	6/10/2009
Gage Serial No.	2345	2345	2345	2345	...	2345	2345	2345
Operator Initials	JKL	RMI	SAU	JKL	...	RMI	SAU	JKL
Sample 1	24.85	25.41	24.98	24.51	...	24.90	24.78	24.70
Sample 2	24.92	24.81	24.84	24.65	...	24.76	25.39	25.31
Sample 3	25.10	24.94	25.24	24.88	...	25.36	25.33	25.34
Sample 4	24.56	24.62	25.26	24.95	...	24.89	25.06	24.81
Sample 5	24.60	25.29	24.63	25.17	...	25.08	24.87	24.82
Sample Set Average = $\bar{X}$	24.806	25.014	24.990	24.832	...	24.998	25.086	24.996
Sample Set Range = R	0.54	0.79	0.63	0.66	...	0.60	0.61	0.64

$$\text{UCL}_R = \mathcal{D}_4 \times \bar{R} \quad \text{and} \quad LCL_R = \mathcal{D}_3 \times \bar{R} \qquad (6.6)$$

where the constants A_2, $\mathcal{D}_3$, and $\mathcal{D}_4$ can be found in A. J. Duncan's *Quality Control and Industrial Statistics* (1974), p. 968, Table M. For this case, namely for a sample size of 5, $A_2 = 0.577$, $\mathcal{D}_3 = 0$, and $\mathcal{D}_4 = 2.115$. For sample sizes 2 through 10, the constants A_2, $\mathcal{D}_3$, and $\mathcal{D}_4$ are also provided in Table 6.2. Control limits are reflections of natural or common variability in our processes. They do not reflect specification limits or objectives.

For the initial $\bar{X}$ and R control chart data for thermal sensor resistance given in Table 6.1,

$$\text{UCL}_{\bar{X}} = 25.256 \text{ Ohms} \quad \text{and} \quad LCL_{\bar{X}} = 24.626 \text{ Ohms} \quad (6.7)$$

$$UCL_R = 1.153 \text{ Ohms} \quad \text{and} \quad LCL_R = 0 \text{ Ohm} \qquad (6.8)$$

$\bar{X}$ and R control charts for the initial 25 sample groups of thermal sensor resistance are given in Figures 6.1 and 6.2.

Figure 6.1 shows sensor resistance averages of five samples per shift. Figure 6.1 also shows $\bar{\bar{X}}$, $\text{UCL}_{\bar{X}}$, and $\text{LCL}_{\bar{X}}$. Figure 6.2 shows sensor resistance ranges of five samples per shift. Figure 6.2 also shows $\bar{R}$, UCL_R, and LCL_R. All averages and ranges per shift are within the

Table 6.2 $\bar{X}$ and R Variable Control Chart Constants A_2, $\mathcal{D}_3$, and $\mathcal{D}_4$

SAMPLE SIZE	A_2	$\mathcal{D}_3$	$\mathcal{D}_4$
2	1.880	0	3.267
3	1.023	0	2.575
4	0.729	0	2.282
5	0.577	0	2.115
6	0.483	0	2.004
7	0.419	0.076	1.924
8	0.373	0.136	1.864
9	0.337	0.184	1.816
10	0.308	0.223	1.777

Note: Factors A, A_2, B_3, B_4, d_2, $\dfrac{1}{d_2}$, d_3, and D_1–D_4 are reproduced with permission from Table B2 of the ASTM Manual on Quality Control of Materials, January 1951, p. 115. Factors A_3, B_5, B_6, and c_4 are reproduced with permission from American Society for Quality (ASQC) Standard A1, Table 1.

For control chart factors, most text books refer to ASTM Manual on Quality Control of Materials, January 1951, p. 115, Table B2.

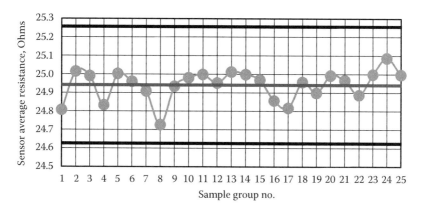

Figure 6.1   Control chart for sensor sample averages per shift in Ohms.

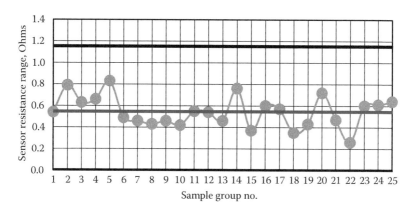

Figure 6.2   Control chart for sensor sample ranges per shift in Ohms.

upper and lower control limits. Most points are close to the means of the control charts. A few points spread out and are close to control limits. It seems that our process is in control during this initial evaluation. However, we do not know if our high-volume manufacturing process is capable at this point in our evaluation to deliver all of our sensors within 25 ± 0.5 Ohms resistance specification to our customer(s). Our processes must be in statistical control before process capability can be determined.

Before we can determine the process capability of our sensors, let us interpret different anomalies that can occur in $\bar{X}$ and R control charts for statistically out-of-control conditions. First, we have to verify that our control charts are in statistical control. An anomaly that is easy to detect for out-of-control conditions is a situation in which one or more

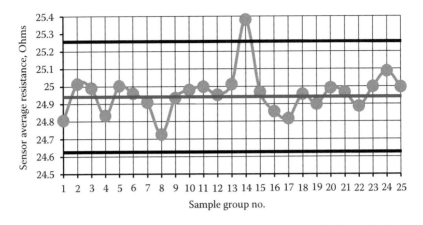

Figure 6.3 Averages control chart out-of-control example in which Sample Group 14 on 6/6/2009 in the afternoon shift was above the upper control limit.

points fall beyond control limits—above $\text{UCL}_{\bar{X}}$ or below $\text{LCL}_{\bar{X}}$—on the averages, $\bar{X}$, chart as shown in the example in Figure 6.3.

This out-of-control point can be caused by

- An unusual process shift during that afternoon
- An error in calculating or plotting the sample averages during that afternoon shift
- A measurement system error (either the gage or the operator)

Since these are the first 25 sample groups from our process control for this sensor, we can eliminate this out-of-control point and recalculate necessary averages and control limits for our initial control charts using 24 sample groups, if we know the special cause for that out-of-control point. If more points are beyond control limits in the initial data set, we have to determine special causes that instigated these out-of-control points. After we complete our detective work and correct for special causes that created out-of-control points, we have to scrap the initial control chart data. We have to take another 25 sample groups to generate fresh initial control charts.

Another anomaly that is easy to detect is the situation in which one or more points can fall above the upper control limit, UCL_R, or below the lower control limit, LCL_R—when $\text{LCL}_R > 0$—on the ranges, R, chart as shown in the example in Figure 6.4.

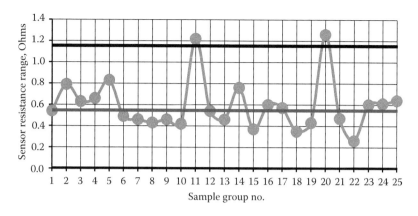

Figure 6.4   Ranges control chart out-of-control example in which Sample Groups 11 and 20 were above the upper control limit.

These out-of-control points can be caused by

- An unusual process shift during those shifts
- An error in calculating or plotting the sample ranges during those shifts
- A measurement system error (either the gage or the operator)

Since more than one data point is beyond control limits in this initial data set, we have to determine special causes that instigated these out-of-control points. After we complete our detective work and correct for special causes that caused out-of-control limits points, we have to scrap the initial control chart data. We have to take another 25 sample groups to create fresh initial control charts.

Other anomalies that will generate statistically out-of-control conditions in $\bar{X}$ and R control charts that we have to watch for are data run patterns and data trend patterns within the control limits. Data runs of eight or more points in a row successively either below $\bar{\bar{X}}$ or above $\bar{\bar{X}}$ in the averages control charts indicate signs of changing processes or measurement system. An example of a data run in an averages control chart is shown in Figure 6.5.

Run patterns in a ranges control chart can mean a greater or smaller spread for the variable that is being investigated. If the run pattern is below the average range the variable spread is smaller. If the run pattern is above the average range the variable spread is larger. A smaller

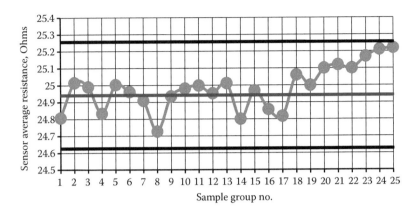

Figure 6.5 Averages control chart out-of-control example for a run in which eight or more points in a row are successively above $\overline{\overline{X}}$.

variable spread run example for a ranges control chart—a run from Sample Groups 5 through 13—is presented in Figure 6.6.

Data trend patterns can be attributed to special variation(s) in our production processes or a change in the measurement system. A data trend has six increasing or decreasing intervals in a row. Typical data trend patterns in control charts are shown in Figures 6.7 and 6.8.

Six or more intervals successively increasing or decreasing data trend patterns shown in Figures 6.7 and 6.8 suggest special causes of variations in our processes or in our measurement system during that data trend period.

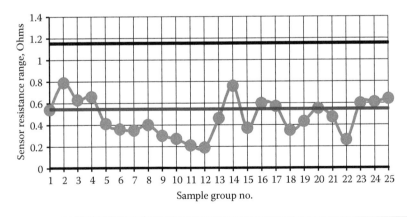

Figure 6.6 Ranges control chart out-of-control example run below the average range, which means smaller sensor resistance spread during that run.

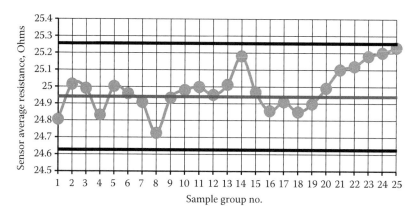

Figure 6.7 An increasing data trend in an averages control chart.

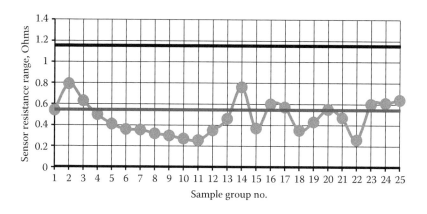

Figure 6.8 A decreasing data trend in a ranges control chart.

There are other nonrandom data patterns in a control chart that should get our attention for out-of-control conditions and require our quick reaction to determine causes for them. These nonrandom patterns can be cyclic, as shown in Figures 6.9 and 6.10.

Another nonrandom data pattern that we should be concerned about occurs when a large quantity of data points—more than two-thirds of data points—lie close to the centerline, $\bar{\bar{X}}$, within a third of the region between control limits—$\bar{\bar{X}} \pm 1\sigma_X$—as shown in Figure 6.11.

Another data pattern case that is out of control when more than two data points in succession lie close to the upper control limit—in a region above $\text{UCL}_{\bar{X}} - 1\sigma_X$ and below $\text{UCL}_{\bar{X}}$, as depicted in

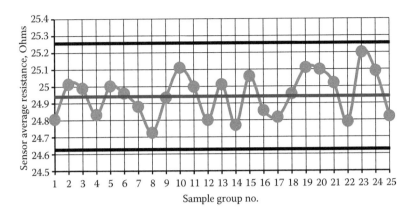

Figure 6.9   Averages control chart example for cyclic nonrandom patterns.

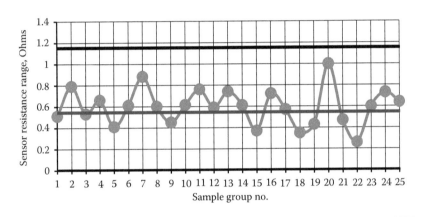

Figure 6.10   Ranges control chart example for cyclic nonrandom patterns.

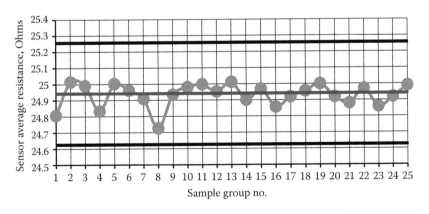

Figure 6.11   Averages control chart example for a nonrandom pattern where more than two-thirds of points are close to the centerline in the $\overline{X} \pm 1\sigma_x$ region.

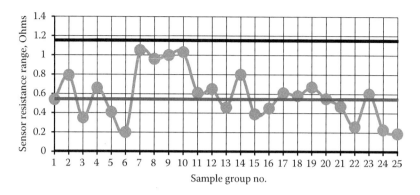

Figure 6.12 Ranges control chart example for a data pattern where more than two data points in succession (data point 7, 8, 9, and 10) are close to the upper control limit.

Figure 6.12. Similar out-of-control data patterns is when more than two data points in succession lie close to lower control limit—in a region above $\mathrm{LCL}_{\bar{X}}$ and below $\mathrm{LCL}_{\bar{X}} + 1\sigma_X$.

Before we can accept that our processes are in statistical control, we have to determine special causes that instigated the following out-of-control cases both in $\bar{X}$ and R control charts.

- Points outside control limits
- Eight points or more data runs in a succession on either side of the average
- Increasing or decreasing data trends successively in six intervals
- Cyclic nonrandom patterns
- Nonrandom patterns close to the centerline (more than two-thirds of points on the control chart)
- Data patterns where more than two data points in succession are close to upper or lower control limits

Once we determine causes for the aforementioned anomalies—out-of-control conditions—in both of our control charts, correct for them, and stabilize our production process, then we can generate a new set of $\bar{X}$ and R control charts that will represent normal causes of variation in our processes. After we have clean and in statistical control charts without any anomalies, we can determine the process capability of our sensor example given in Table 6.1 and plotted in Figures 6.1 and 6.2.

When the initial 25 or so sample points are in statistical control, we can extend the control charts' averages and upper and lower control limits on our averages and ranges charts and start using these two control charts—process average, $\bar{X}$, and the process range, R—as our process control guides in production. We also have to set up strict guidelines as to what to do—reaction plans—when sample data point(s) go out of control and when we start seeing anomalies in our sample data. An operator should be authorized to stop production and immediately inform responsible engineers and managers to solve unexpected process issues when out-of-control sample data point(s) start to appear in our control charts.

Let us go through another $\bar{X}$ and R control charting example. The control chart data is for the plating thickness variable for a towel hanger. The control charting sample size per day is 3. The initial data are for 30 days. Raw data are presented in Table 6.3.

The average of three samples for Sample Group 1 is obtained as follows:

$$\bar{X}_1 = (2.984 + 3.021 + 3.029) / 3 = 3.011 \tag{6.9}$$

The range of five samples for Sample Group 1 is obtained as follows:

$$R_1 = X_{\text{High for Sample Group 1}} - X_{\text{Low for Sample Group 1}}$$
$$= 3.029 - 2.984 = 0.045 \tag{6.10}$$

Then we calculate the process average, $\bar{\bar{X}}$, using Equation 6.3 and the process range, $\bar{R}$, using Equation 6.4 for all sample groups where j is the number of sample groups. For the example given in Table 6.3, $j = 30$.

For the process data in Table 6.3, $\bar{\bar{X}} = 2.988$ μm and $\bar{R} = 0.068$ μm. Next we have to determine control limits for the average chart and for the range chart. Calculations for control limits in $\bar{X}$ and R variable charts use constants and these constants vary according to the sample size. Upper control limits (UCLs) and lower control limits (LCLs) in $\bar{X}$ and R variable charts are defined in Equations 6.5 and 6.6, where the constants A_2, D_3, and D_4 can be found in A. J. Duncan's *Quality Control and Industrial Statistics* (1974), p. 968, Table M. For this case, namely for a sample size of 3, $A_2 = 1.023$, $D_3 = 0$, and $D_4 = 2.575$. For sample sizes 2 through 10, constants A_2, D_3, and D_4 are also provided in Table 6.2. I would like to reemphasize that control limits are

Table 6.3 Initial $\bar{X}$ and R Control Chart Data for Towel Hanger Plating Thickness

PRODUCT NAME AND PART NO.: TOWEL HANGER TH-165	PROCESS: PLATING			VARIABLE: PLATING THICKNESS IN MICROMETERS	SPECIFICATION LIMITS = 3.000 ± 0.100 MMETERS	SAMPLE SIZE = 3 SAMPLE FREQUENCY = EVERY TWO HOURS		
Sample Group No.	1	2	3	4		28	29	30
Sample Time	10 AM, 1 PM, 4PM	10 AM, 1 PM, 4PM	10 AM, 1 PM, 4PM	10 AM, 1 PM, 4PM	...	10 AM, 1 PM, 4PM	10 AM, 1 PM, 4PM	10 AM, 1 PM, 4PM
Sample Date	8/2/2006	8/3/2006	8/4/2006	8/5/2006	...	8/29/2006	8/30/2006	8/31/2006
Gage Serial No.	16A	16A	16A	16A	...	16A	16A	16A
Operator Initials	GKA	GKA	GKA	CAV	...	GKA	GKA	GKA
Sample 1	2.984	3.066	3.046	3.004	...	2.908	2.915	2.913
Sample 2	3.021	2.921	3.057	2.979	...	2.949	2.907	2.952
Sample 3	3.029	2.908	3.078	2.981	...	2.974	2.979	2.936
Sample Set Average = $\bar{X}$	3.011	2.965	3.060	2.988	...	2.944	2.934	2.934
Sample Set Range = R	0.045	0.158	0.032	0.025	...	0.066	0.072	0.039

reflections of natural or common variability in our processes. They do not reflect specification limits or objectives.

For the initial $\bar{X}$ and R control chart data for plating thickness given in Table 6.3,

$$\text{UCL}_{\bar{X}} = 3.058\,\mu\text{m} \quad \text{and} \quad \text{LCL}_{\bar{X}} = 2.919\,\mu\text{m} \qquad (6.11)$$

$$\text{UCL}_R = 0.175\,\mu\text{m} \quad \text{and} \quad \text{LCL}_R = 0\,\mu\text{m} \qquad (6.12)$$

$\bar{X}$ and R control charts for the initial 30 sample groups of plating thickness are given in Figures 6.13 and 6.14.

First we have to check if our control charts are in statistical control for the initial 30 days. We analyze sample data for out-of-control conditions both in $\bar{X}$ and R control charts like we did for the first example, namely for

- Points outside control limits
- Eight points or more data runs in a succession on either side of the average
- Increasing or decreasing data trends successively in six intervals
- Cyclic nonrandom patterns
- Nonrandom patterns close to the centerline (more than two-thirds of points on the control chart)
- Data patterns where more than two data points in succession are close to upper or lower control limits

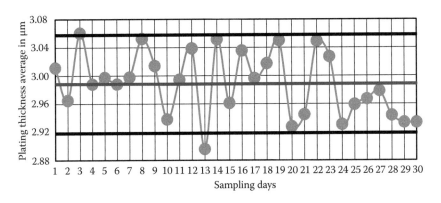

Figure 6.13 Control chart for towel hanger plating thickness sample averages per day in micrometers.

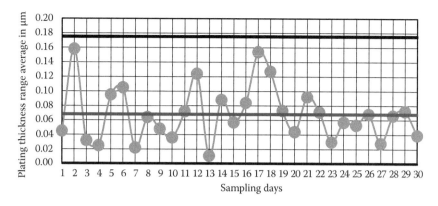

Figure 6.14 Control chart for towel hanger plating thickness sample ranges per day in micrometers.

We see that the sample averages data point for the 13th day is below the lower control limit in Figure 6.13. The cause for this out-of-control data point was recorded by the test operator as "misread data." If we eliminate the sample average data point for day 13, $\bar{X}$ and R control charts look clean from any out-of-control conditions. We can now determine the process capability (see Chapter 10 for details of process capability) of our plating thickness process.

Process standard deviation can be estimated from the mean of sample ranges by using Equation 10.1. However, we have to recalculate the average of ranges by excluding the out-of-control data point belonging to day 13. The recalculated average range is $\bar{R} = 0.070$ μm and

$$\sigma = \frac{\bar{R}}{d_2} = \frac{0.070}{1.693} = 0.041 \ \mu m \tag{6.13}$$

where d_2 depends on sample size and it can be found in A. J. Duncan's *Quality Control and Industrial Statistics* (1974), p. 968, Table M. For a sample size of 3, $d_2 = 1.693$. The constants d_2 for different sample sizes are also given in Table 10.1.

If we want a 6σ capability for our plating thickness process for our customer's ship-to-stock requirement, which is a 4.5σ process standard deviation with allowance for the 1.5σ process average shift, we have to reduce our plating thickness process spread to

$$-\frac{(\text{Lower Specification Limit} - \bar{\bar{X}})}{4.5} = -\frac{(2.900 - 2.991)}{4.5} = 0.020 \ \mu m.$$

The new $\bar{\bar{X}} = 2.991\,\mu m$ value was obtained by excluding the out-of-control data point belonging to day 13. This result shows that we have to halve the spread of our plating thickness processes to meet our customer's 6σ requirement assuming that our plating thickness average, $\bar{\bar{X}}$, will not shift.

We form a team and upgrade our plating processes. Uniformity and cleanliness of our electrolyte bath were improved to very high levels. After stabilizing our improved plating processes, we have to take control chart samples for 30 days. $\bar{X}$ and R control charts for the improved plating process are shown in Figures 6.15 and 6.16.

For improved plating process data, $\bar{\bar{X}} = 2.996\,\mu m$ and $\bar{R} = 0.032\,\mu m$. The upper and lower control limits for the improved plating process are as follows:

$$\text{UCL}_{\bar{X}} = 3.028\,\mu m \quad \text{and} \quad \text{LCL}_{\bar{X}} = 2.963\,\mu m \qquad (6.14)$$

$$\text{UCL}_R = 0.083\,\mu m \quad \text{and} \quad \text{LCL}_R = 0\,\mu m \qquad (6.15)$$

$\bar{X}$ and R control chart limits reflect that common variability in our improved plating process reduced quite a bit.

Then we check if our $\bar{X}$ and R control charts are in statistical control for the initial 30 days. In Figures 6.15 and 6.16 all points and trends seem to be in statistical control. Now we can estimate

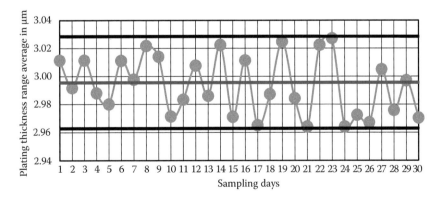

Figure 6.15 Control chart for towel hanger plating thickness from improved process sample averages per day in micrometers.

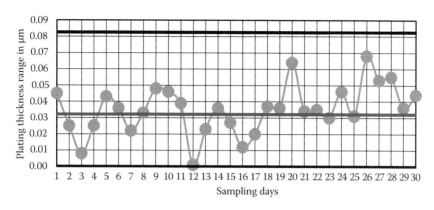

Figure 6.16 Control chart for towel hanger plating thickness from improved process sample ranges per day in micrometers.

the standard deviation for our improved plating process by using Equation 10.1.

$$\sigma = \frac{\bar{R}}{d_2} = \frac{0.032}{1.693} = 0.019 \ \mu m \qquad (6.16)$$

For customer qualification, we had to reduce our improved plating thickness processes' standard deviation to

$$-\frac{\left(\text{Lower Specification Limit} - \bar{\bar{X}}\right)}{4.5} = -\frac{\left(2.900 - 2.996\right)}{4.5} = 0.021 \ \mu m.$$

We improved the required plating thickness process spread by $\left(\dfrac{0.021 - 0.019}{0.021}\right) \times 100 = 9.5\%$.

7

PROCESS CONTROL FOR VARIABLES: $\bar{X}$ AND S CHARTS

Other important control charts for variables are $\bar{X}$ and S charts. $\bar{X}$ and S charts should be used when the spread of a variable is critical. If your process capability index is close to one and you are not 100% sorting for a particular variable, you should use $\bar{X}$ and S charts to determine the standard deviation of your processes for that particular variable more accurately. The minimum sample size for $\bar{X}$ and S charts should be 10, but sample size can go up to the 100's depending on your production volume and on your automated voluminous data capturing and analysis capability. $\bar{X}$ and S charts are treated and analyzed similarly to $\bar{X}$ and R charts in Chapter 6, but instead of ranges, standard deviations are utilized in estimating the process spread. Also, different constants are used in determining upper and lower control limits for $\bar{X}$ and S charts.

The average of a sample group, $\bar{X}$, is obtained like in $\bar{X}$ and R charts in Chapter 6 as follows:

$$\bar{X} = \frac{1}{N} \sum_{i=1}^{N} X \tag{7.1}$$

where N is the number of samples in a group.

Then we calculate the process average, $\bar{\bar{X}}$, over all sample groups. If we identify j as the number of sample groups, we get Equation 7.2.

$$\bar{\bar{X}} = \frac{(\bar{X}_1 + \bar{X}_2 + \bar{X}_3 + \cdots + \bar{X}_{j-1} + \bar{X}_j)}{j} \tag{7.2}$$

Then we calculate the standard deviation for each sample group as follows:

$$S = \sqrt{\frac{\sum_{i=1}^{N}(X - \bar{X})^2}{N-1}} \tag{7.3}$$

Then we calculate the process average, $\bar{S}$, over all sample groups as follows:

$$\bar{S} = \frac{(S_1 + S_2 + S_3 + \cdots + S_{j-1} + S_j)}{j} \tag{7.4}$$

Next we have to determine control limits for the averages chart and for the standard deviations chart. Calculations for control limits in $\bar{X}$ and S variable charts use constants and these constants vary according to the sample size. Upper control limits (UCLs) and lower control limits (LCLs) in $\bar{X}$ and S variable charts are defined as follows:

$$\text{UCL}_{\bar{X}} = \bar{\bar{X}} + A_3 \times \bar{R} \quad \text{and} \quad \text{LCL}_{\bar{X}} = \bar{\bar{X}} - A_3 \times \bar{R} \tag{7.5}$$

$$\text{UCL}_S = B_4 \times \bar{S} \quad \text{and} \quad \text{LCL}_S = B_3 \times \bar{S} \tag{7.6}$$

where the constants A_3, B_3, and B_4 can be found in A. J. Duncan's *Quality Control and Industrial Statistics* (1974), p. 968, Table M. For a case that will be studied in the text that follows with a sample group size of 25, $A_3 = 0.606$, $B_3 = 0.565$, and $B_4 = 1.435$. For sample group sizes 10 through 25, the constants A_3, B_3, and B_4 are also provided in Table 7.1. We should again emphasize that control limits are reflections of natural or common variability in our processes. They do not reflect specification limits or objectives.

The following $\bar{X}$ and S variable control chart example is for X dimension alignment of a magnetic sensor on a suspension. The customer specification for the X dimension alignment is 5000 ± 80 µm. Our processes capability index for this variable is very close to one. We want to monitor closely our processes' spread for this variable. We collect 25 samples at random out of every day's production of about 1000 sensor assemblies. Control chart sample data for the first 25 days are provided in Table 7.2.

Table 7.1 $\bar{X}$ and S Variable Control Charts' Limits Constants A_3, B_3, and B_4

SAMPLE SIZE	A_3	B_3	B_4
10	0.975	0.284	1.716
11	0.927	0.321	1.679
12	0.886	0.354	1.646
13	0.850	0.382	1.618
14	0.817	0.406	1.594
15	0.789	0.428	1.572
16	0.763	0.448	1.552
17	0.739	0.466	1.534
18	0.718	0.482	1.518
19	0.698	0.497	1.503
20	0.680	0.510	1.490
21	0.663	0.523	1.477
22	0.647	0.534	1.466
23	0.633	0.545	1.455
24	0.619	0.555	1.445
25	0.606	0.565	1.435

Note: Factors A, A_2, B_3, B_4, d_2, $\dfrac{1}{d_2}$, d_3, and D_1–D_4 are reproduced with permission from Table B2 of the ASTM Manual on Quality Control of Materials, January 1951, p. 115. Factors A_3, B_5, B_6, and c_4 are reproduced with permission from American Society for Quality (ASQC) Standard A1, Table 1.

For control chart factors, most text books refer to ASTM Manual on Quality Control of Materials, January 1951, p. 115, Table B2.

The average of 25 samples for Sample Group 1 is obtained as follows:

$$\bar{X}_1 = \frac{(5034.8 + 4975.7 + 4970.6 + \cdots + 5017.0 + 4987.8 + 5022.8)}{25}$$

$$= 4998.1 \ \mu m \tag{7.7}$$

The standard deviation of 25 samples for Sample Group 1 is obtained using Equation 7.3 as follows:

$$S_1 = \sqrt{\frac{(5034.8 - 4998.1)^2 + (4975.7 - 4998.1)^2 + \cdots + (4987.8 - 4998.1)^2 + (5022.8 - 4998.1)^2}{(25 - 1)}}$$

$$= 18.75 \ \mu m$$

$$\tag{7.8}$$

Then we calculate the process average, $\bar{\bar{X}}$, and the process standard deviation, $\bar{S}$, for all 25 sample groups using Equations 7.2 and 7.4. For the process data in Table 7.2, $\bar{\bar{X}} = 4999.6 \ \mu m$ and $\bar{S} = 22.71 \ \mu m$.

Table 7.2 Initial $\bar{X}$ and S Control Chart Data for Magnetic Sensor–Suspension X Dimension Alignment

PRODUCT NAME AND PART NO.: MAGNETIC SENSOR 10753-01	PROCESS: SUSPENSION ALIGNMENT			VARIABLE: X-DIMENSION ALIGNMENT IN MICROMETERS		SPECIFICATION LIMITS = 5000 ± 80 μm	SAMPLE SIZE = 25 SAMPLE FREQUENCY = AT RANDOM 3 SAMPLES PER HOUR		
Sample Group #	1	2	3	4		23	24	25
Sample Time	9 AM	9 AM	9 AM	9 AM		9 AM	9 AM	9 AM
Sample Date	3/1/2009	3/2/2009	3/3/2009	3/4/2009		3/23/2009	3/24/2009	3/25/2009
Gage Serial No.	1195-S	1195-S	1195-S	1195-S		1195-S	1195-S	1195-S
Operator Initials	EOA	EOA	GKB	GKB		GKB	EOA	EOA
Sample 1	5034.8	5012.1	5029.3	4983.8	⋮	5032.3	5034.6	4982.8
Sample 2	4975.7	4982.3	5016.8	5015.0	⋮	5012.2	5030.4	4970.3
Sample 3	4970.6	4978.1	4997.5	4980.3	⋮	5007.8	5032.3	4966.6
Sample 4	4985.7	5026.4	4964.1	4979.0	⋮	4968.6	4984.7	4999.5
Sample 5	4985.0	5036.9	5011.9	4982.3	⋮	4982.9	5036.4	4981.3
⋮								
Sample 21	4999.5	4962.8	5033.8	4999.5		5016.6	4977.9	4996.0
Sample 22	4983.9	5032.8	4971.1	4983.9		4982.3	4973.4	5025.3
Sample 23	5017.0	4968.6	4999.3	5017.0		4988.9	5038.9	4970.5
Sample 24	4987.8	4966.9	4964.5	4987.8		5005.6	5033.2	4966.4
Sample 25	5022.8	4963.2	5029.1	5022.8		4972.3	4960.3	4994.7
Sample Set Average = $\bar{X}$	4998.1	4994.7	5003.7	4989.2	⋮	5004.5	5004.6	4996.1
Sample Set Standard Deviation = S	18.75	24.99	26.37	19.03		23.15	29.77	22.16

Note: Factors A, A_2, B_3, B_4, d_2, $\frac{1}{d_2}$, d_3, and D_1–D_4 are reproduced with permission from Table B2 of the ASTM Manual on Quality Control of Materials, January 1951, p. 115. Factors A_3, B_5, B_6, and c_4 are reproduced with permission from American Society for Quality (ASQC) Standard A1, Table 1. For control chart factors, most text books refer to ASTM Manual on Quality Control of Materials, January 1951, p. 115, Table B2.

Next we have to determine control limits for the averages chart and for the standard deviations chart. Calculations for control limits in $\bar{X}$ and S variable charts use constants and these constants vary according to the sample size as shown in the preceding text. Upper control limits (UCLs) and lower control limits (LCLs) in $\bar{X}$ and S variable charts for this example are as follows:

$$\text{UCL}_{\bar{X}} = 4999.6 + 0.606 \times 22.71 = 5013.4 \ \mu\text{m} \ \text{ and}$$
$$\text{LCL}_{\bar{X}} = 4999.6 - 0.606 \times 22.71 = 4985.8 \ \mu\text{m} \qquad (7.9)$$

$$\text{UCL}_{S} = 1.435 \times 22.71 = 32.59 \ \mu\text{m and}$$
$$\text{LCL}_{S} = 0.565 \times 22.71 = 12.83 \ \mu\text{m} \qquad (7.10)$$

Plots of $\bar{X}$ and S variable charts for this example are shown in Figures 7.1 and 7.2.

Before we can accept that our sensors' X dimension alignment processes are stable, we have to determine if the aforementioned $\bar{X}$ and S variable charts for the initial 25 days of production are in statistical control or not. Statistical out-of-control checks for $\bar{X}$ and S control charts are the same as the ones presented in Chapter 6 for $\bar{X}$ and R control charts, namely

- Points outside control limits
- Eight points or more data runs in a succession on either side of the average
- Increasing or decreasing data trends successively in six intervals

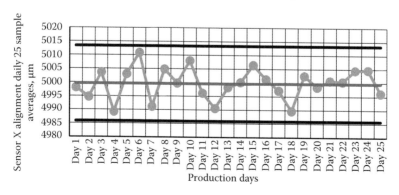

Figure 7.1 Sensor X dimension alignment daily 25 sample averages for 25 days along with $\bar{\bar{X}}$ and upper and lower control limits.

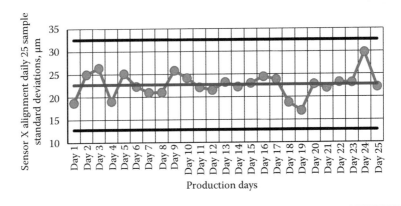

Figure 7.2 Sensor X dimension alignment daily 25 sample standard deviations for 25 days along with $\bar{S}$ and upper and lower control limits.

- Cyclic nonrandom patterns
- Nonrandom patterns close to the centerline (more than two-thirds of points on the control chart)
- Data patterns where more than two data points in succession are close to the upper or lower control limits

The $\bar{X}$ and S charts in Figures 7.1 and 7.2 do not have any points outside control limits. They do not show any runs, namely eight or more points in a row on either side of the average. They also do not show increasing or decreasing data trends at six intervals in succession. Both $\bar{X}$ and S charts do not show any cyclic nonrandom patterns that might be caused by special causes of variation in our processes. Both $\bar{X}$ and S charts do not show any data patterns close to the centerline or close to control limits. We can conclude that our sensors' X dimension alignment processes are stable and in statistical control. We can now go ahead and determine our processes' standard deviation for this variable.

If one of our initial control charts is out of control, we have to determine special causes that instigated these out-of-control conditions. After we complete our detective work and correct for special causes that created the out-of-control conditions, we have eliminated out-of-control data points or we have to scrap the initial control chart data. If the whole data set is scrapped, then we have to take another 25 sample groups to generate fresh initial $\bar{X}$ and S charts until we get our processes in control for this variable.

Any one of the control charts can go out of control at any time during a production process. We have to have strict procedures in place as to what to do and how to deal with out-of-control conditions, including shutting down production and calling emergency meetings. The people who are involved all the way to the top management of the company should be clearly identified for every control chart.

Next we can estimate the process standard deviation for this variable, since our $\bar{X}$ and S charts are in control. We have to use Equation 7.11 to estimate the sensor X dimension alignment standard deviation.

$$\sigma_{process} = \frac{\bar{S}}{c_4} \tag{7.11}$$

where the constants c_4 can be found in A. J. Duncan's *Quality Control and Industrial Statistics* (1974), p. 968, Table M and provided in Table 7.3.

Table 7.3 Process Standard Deviation Constant c_4 in Equation 7.11

SAMPLE SIZE	c_4
10	0.9727
11	0.9754
12	0.9776
13	0.9794
14	0.9810
15	0.9823
16	0.9835
17	0.9845
18	0.9854
19	0.9862
20	0.9869
21	0.9876
22	0.9882
23	0.9887
24	0.9892
25	0.9896

Note: Factors A, A_2, B_3, B_4, d_2, $\frac{1}{d_2}$, d_3, and D_1–D_4 are reproduced with permission from Table B2 of the ASTM Manual on Quality Control of Materials, January 1951, p. 115. Factors A_3, B_5, B_6, and c_4 are reproduced with permission from American Society for Quality (ASQC) Standard A1, Table 1.

For control chart factors, most text books refer to ASTM Manual on Quality Control of Materials, January 1951, p. 115, Table B2.

Our estimated process standard deviation for sensor X dimension alignment is

$$\sigma_{\text{Sensor X alignment}} = \frac{22.71}{0.9896} = 22.95 \ \mu m \qquad (7.12)$$

We can also calculate the process standard deviation from the initial 625 data points—25 samples every day for 25 days—and it comes out to be 23.11 μm. There is only a 0.7% difference between our estimated standard deviation value from Equation 7.11 and our large data base. Therefore Equation 7.12 represents the spread of our processes very well for this variable.

Since our $\bar{X}$ and S charts are in control, we can calculate our process capability by assuming a normal distribution; see Chapter 10. By using our customer's sensor X dimension alignment specification—5000 ± 80 μm, the sensor X dimension alignment process mean—$\bar{\bar{X}} = 4999.59$ μm, the sensor X dimension alignment process standard deviation $\sigma = 22.95$ μm, we determine the nondimensional upper specification point, Z_{USL}, as follows:

$$Z_{\text{USL}} = \frac{(\text{Upper Specification Limit} - \bar{\bar{X}})}{\sigma} = 3.504 \qquad (7.13)$$

and we determine the nondimensional lower specification point, Z_{LSL}, as follows:

$$Z_{\text{LSL}} = \frac{(\text{Lower Specification Limit} - \bar{\bar{X}})}{\sigma} = -3.469 \qquad (7.14)$$

Our process capability index, C_{pk}, is the minimum absolute value in Equations 7.13 and 7.14.

$$C_{\text{pk}} = \frac{Z_{\text{min}}}{3} = \frac{3.469}{3} = 1.156 \qquad (7.15)$$

We can next calculate the number of sensors that will be out of specification for X dimension alignment using MS Excel®, namely above 5080 μm.

$$1 - \text{NORMALDIST}(5080, 4999.59, 22.71, \text{TRUE})$$
$$= 0.000229 \text{ or } 0.0229\% \tag{7.16}$$

We will have 229 rejects out of a million above the 5080 µm specification.

The percentage of parts falling below the X dimension alignment specification of 4920 µm is

$$\text{NORMALDIST}(4920, 4999.59, 22.71, \text{TRUE})$$
$$= 0.000262 \text{ or } 0.0262\% \tag{7.17}$$

We will have 262 rejects out of a million below the 4920 µm specification. Our total rejects out of a million for the sensor X dimension alignment specification of 5000 ± 80 µm will be 229 + 262 = 491.

If our customer requires a 6σ process capability for the sensor X dimension alignment, we have to reduce the standard deviation of our sensor X dimension alignment processes to

$\sigma_{6\sigma} = \dfrac{(5080 - 5000)}{6} = 13.33$ µm. This improvement in spread requires

a 41.3%—$\dfrac{(22.71 - 13.33)}{22.71} \times 100 = 41.3\%$—reduction in standard deviation for our X alignment processes. Remember that ±1.5σ out of ±5σ spread is allocated to mean, $\bar{\bar{X}}$, shift in our X alignment processes in the 6σ process capability definition. On the other hand, if we can get our customer to agree to widen the specification for X alignment to 5000 ± 100 µm, then our present process capability will be very close

to 4.5σ, namely $Z = \dfrac{(4900 - 4999.59)}{22.71} = -4.39$ below the mean and

$Z = \dfrac{(5100 - 4999.59)}{22.71} = +4.42$ above the mean.

Let us perform another $\bar{X}$ and S variable control chart example. This example is for qualification of a new photoresist for our 6-inch wafer line. This new photoresist is supposed to give us much more control in forming a critical feature on our sensors. The specification for this critical feature is 1.000 ± 0.090 µm. We have initially performed several designs of experiments to optimize our processes' characteristics with this new photoresist. After the optimum processes' characteristics were determined, we ran 25 wafers using the new photoresist. Then we measured 25 sensors' critical feature at random on every wafer. Sample data obtained from these measurements are presented in Table 7.4.

Table 7.4 New Photoresist Qualification Sample Data for the Proximity Sensor's Critical Feature, $\bar{\bar{X}} = 1.016$ μm, and the Process Standard Deviation, $\bar{S} = 0.0209$ μm.

PRODUCT NAME AND PART NO.: PROXIMITY SENSOR JHS-041	PROCESS: CRITICAL FEATURE PHOTOLITHOGRAPHY		VARIABLE: CRITICAL FEATURE DIMENSION IN MICROMETERS		SPECIFICATION LIMITS = 1.000 ± 0.090 μm	SAMPLE SIZE = 25 PER WAFER AT RANDOM		
Wafer No.	1	2	3	4		23	24	25
Time	10 AM	1 PM	3 PM	5 PM	⋮	9 AM	11 AM	2 PM
Sample Date	11/10/2009	11/10/2009	11/10/2009	11/10/2009	⋮	11/11/2009	11/11/2009	11/11/2009
Gage No.	21-K	21-K	21-K	21-K	⋮	21-K	21-K	21-K
Operator	GKZ	GKZ	GKZ	GKZ	⋮	GKZ	GKZ	GKZ
Sample 1	0.985	0.986	0.989	0.993	⋮	1.001	0.997	1.001
Sample 2	0.981	0.988	0.989	0.991	⋮	1.019	1.015	1.019
Sample 3	0.988	0.991	0.990	0.994	⋮	1.016	1.012	1.016
Sample 4	0.993	0.999	1.001	1.007	⋮	1.003	1.011	1.013
Sample 5	1.002	1.002	1.006	1.010	⋮	1.057	1.021	1.037
⋮	⋮	⋮	⋮	⋮	⋮	⋮	⋮	⋮
Sample 21	1.055	1.056	1.050	1.054	⋮	1.021	1.020	1.001
Sample 22	0.977	0.982	0.990	0.994	⋮	1.031	1.032	1.012
Sample 23	0.972	0.994	0.996	0.985	⋮	0.935	0.937	0.945
Sample 24	0.991	0.996	0.997	1.001	⋮	1.027	1.028	1.010
Sample 25	0.968	0.995	0.973	0.971	⋮	1.007	1.005	1.002
25 Sample Average = $\bar{X}$	1.006	1.012	1.011	1.015	⋮	1.020	1.016	1.015
25 Sample Standard Deviation = S	0.022	0.020	0.019	0.021	⋮	0.025	0.021	0.020

With the old photoresist, the proximity sensor's critical feature had a normal distribution with a process average, $\bar{\bar{X}} = 1.010$ μm, and the process standard deviation, $\bar{S} = 0.025$ μm.

Since in our old processes $\bar{X}$ and S charts were in control and our data showed a normal distribution behavior, we can calculate our process capability for this critical feature; see Chapter 10. By using the proximity sensor's critical feature specification—1.000 ± 0.090 μm— we can determine the nondimensional upper specification point, Z_{USL}, as follows:

$$Z_{USL} = \frac{(1.090 - 1.010)}{0.025} = 3.2 \qquad (7.18)$$

and we determine the nondimensional lower specification point, Z_{LSL}, as follows:

$$Z_{LSL} = \frac{(0.910 - 1.010)}{0.025} = -4.0 \qquad (7.19)$$

Our process capability index, C_{pk}, is minimum absolute value in Equations 7.18 and 7.19.

$$C_{pk} = \frac{Z_{min}}{3} = \frac{3.2}{3} = 1.07 \qquad (7.20)$$

We can next calculate the number of proximity sensors that were out of the high end specification for this critical feature when we used the old photoresist, namely above 1.090 μm.

$$1 - NORMALDIST(1.090, 1.010, 0.025, TRUE)$$
$$= 0.000687 \text{ or } 0.0687\% \qquad (7.21)$$

We had 687 rejects out of a million above the 1.090 μm specification.

The percentage of parts falling below this critical feature specification of 0.910 μm was

$$NORMALDIST(0.910, 1.010, 0.025, TRUE)$$
$$= 0.000032 \text{ or } 0.0032\% \qquad (7.22)$$

We will have 32 rejects out of a million below the 0.910 μm specification. Our total rejects out of a million proximity sensors for this critical feature specification of 1.000 ± 0.090 μm will be 687 + 32 = 719.

Now let us investigate to see if the new photoresist that we are trying to qualify improves the process capability for this critical feature. The data shown in Table 7.4 should first be tested for normality. Figure 7.3 for wafer 1 data and Figure 7.4 for the data of all 25 wafers are plotted to compare frequency and normal distributions.

Figures 7.3 and 7.4 show that individual wafer data and data for all 25 wafers exhibit a normal behavior. Next we have to verify if our new optimized photolithography processes using the new photoresist is under statistical control or not. To observe this behavior, we can plot $\bar{X}$ and S control charts for 25 wafers, shown in Figures 7.5 and 7.6. The upper control limits (UCLs) and lower control limits (LCLs) in $\bar{X}$ and S variable charts are obtained by using Equations 7.5 and 7.6.

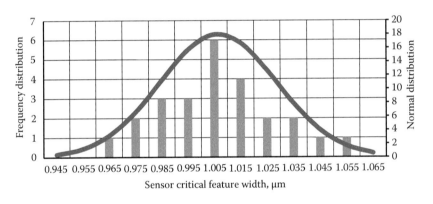

Figure 7.3 Proximity sensor critical feature frequency and normal distributions for wafer 1 obtained when using new photoresist processes.

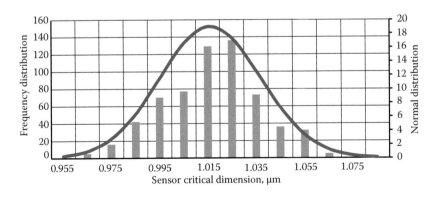

Figure 7.4 Proximity sensor critical feature frequency and normal distributions for 25 wafers obtained when using new photoresist processes.

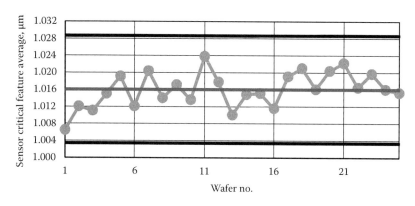

Figure 7.5   Proximity sensor critical feature wafer averages obtained when using new photoresist processes.

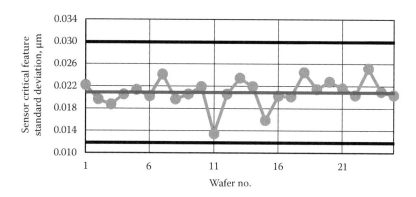

Figure 7.6   Proximity sensor critical feature wafer standard deviations obtained when using new photoresist processes.

Statistical out-of-control checks for these control charts are the same as the ones presented before. We see that both $\bar{X}$ and S control charts for 25 wafers are in statistical control. We can now go ahead and determine the standard deviation and then the process capability for this critical feature resulting from our new photoresist processes. Using Equation 7.11, our new processes' standard deviation is

$$\sigma_{\text{New photoresist process}} = \frac{0.0209}{0.9896} = 0.021 \ \mu\text{m} \tag{7.23}$$

Let us see how much our spread has improved with our new photoresist processes.

New photoresist processes spread improvement

$$= \frac{(0.025 - 0.021)}{0.025} \times 100 = 16\% \tag{7.24}$$

We can now determine the nondimensional upper specification point, Z_{USL}, for our new photoresist processes using $\overline{\overline{X}} = 1.016$ μm and $\overline{S} = 0.021$ μm.

$$Z_{USL} = \frac{(1.090 - 1.016)}{0.021} = 3.52 \tag{7.25}$$

and we determine the nondimensional lower specification point, Z_{LSL}, as follows:

$$Z_{LSL} = \frac{(0.910 - 1.016)}{0.021} = -5.05 \tag{7.26}$$

Our new process capability index, C_{pk}, is the minimum absolute value in Equations 7.25 and 7.26.

$$C_{pk} = \frac{Z_{min}}{3} = \frac{3.52}{3} = 1.17 \tag{7.27}$$

We can next calculate the number of proximity sensors that were out of the high end specification for this critical feature when we used the new photoresist, namely above 1.090 μm.

$$1 - \text{NORMALDIST}(1.090, 1.016, 0.021, \text{TRUE})$$
$$= 0.000213 \text{ or } 0.0213\% \tag{7.28}$$

We had 213 rejects out of a million above the 1.090 μm specification.

The percentage of parts falling below this critical feature specification of 0.910 μm was

$$\text{NORMALDIST}(0.910, 1.016, 0.021, \text{TRUE})$$
$$= 0.00000022 \text{ or } 0.000022\% \tag{7.29}$$

We will have zero rejects out of a million below the 0.910 μm specification. Our total rejects out of a million proximity sensors for this critical feature specification of 1.000 ± 0.090 μm will be 213 + 0 = 213. The new photoresist process for our proximity sensor's critical feature improved the spread noticeably by 16%. However, the new process average also increased from the old 1.010 μm to 1.016 μm. If we can fine-tune our new processes and lower the process average toward 1.010 μm, we will have almost no rejects for this critical feature.

8

PROCESS CONTROL FOR ATTRIBUTES

P, NP, C, and U Charts

Attribute type data has noncontinuous values, namely "go" or "no-go," "conforms" or "nonconforms," "pass" or "fail," "on time" or "not on time," "small, medium, or large," and so forth. For some inspections such as contamination, scratches, chip, and so forth, it might be difficult to define the difference between a conforming and a nonconforming or defective product unit, especially when an inspector's judgment is involved. Precise operational specifications for attribute characteristics are a must. Thorough training of inspectors is also required. Attribute data versus time can be collected and analyzed using different types of attribute control charts. There are four main types of control charts—P, NP, C, and U charts—used for analysis of attribute data.

P charts are used for proportions of nonconforming or defective product units found in a sample lot. For example, in a lightbulb manufacturing plant during final inspection, what percentage of lightbulbs did not light up from a randomly selected production sample lot? Another P chart example can be for proportion of sales orders with errors to total number of sales orders processed in a week. One more P-chart example can be for proportion of airplanes that arrived late in a 24-hour day to total number of airplanes arrived. A product unit can be inspected for one or more attribute characteristics during an inspection. Sample lot sizes for P charts can be a constant or can vary.

NP charts are used for the number of nonconforming or defective product units (not proportions like in P charts) found in a sample lot. We can also chart the preceding examples for P charts by plotting the number of nonconforming or defective units versus time. Again, a product unit can be evaluated for one or more attribute characteristics. Sample lot sizes for NP charts must be a constant.

C charts focus on the number of attribute defect(s) instead of nonconforming or defective product units. C charts are used for plotting total number of attribute defect(s) in a sample lot. For example, number of errors on an invoice, number of contaminated regions on a wafer, or number of complaints a hotel receives each day can be controlled by using C charts. A sample lot can be inspected for one or more attribute nonconformities or defects during inspection. Sample lot sizes for C charts are always a constant.

U charts are similar to C charts and they are used for total number of attribute nonconformities or defects per unit sample lot. A sample lot can be inspected for one or more attribute nonconformities or defects during inspection. Sample lot sizes for U charts can vary in size.

P charts require large sample sizes. Also, sample sizes should be a constant, if possible. If sample sizes cannot be a constant, then they should not vary by more than ±25%. Before collecting attribute data, conforming and nonconforming definitions should be clearly established and everyone involved with these control charts should be well trained. Initially 25 sample lots of attribute data should be collected and control charted including the process average and upper and lower control limits. Finally, initial P charts should be analyzed for in control or out-of-control criteria. If a sample size varies more than ±25% from the standard and agreed on sample size, then upper and lower control limits for that particular sample should be recalculated.

The inspector has to record the number of nonconforming or defective products from a sample lot on to an attribute data sheet as shown in Table 8.1. The inspector rejects a sample product for contamination, edge chip, crack, and lead defect specifications.

The next to the last row in Table 8.1 shows the calculated proportion of nonconforming or defective sample products relative to the number of inspected products for each sample lot.

Proportion of defective products to sample size = NP/N (8.1)

Then the last row in Table 8.1 shows the percentage of nonconforming or defective sample products relative to the number of inspected products for each sample lot. These percentage values are obtained by multiplying the proportion values from Equation 8.1 by 100.

Table 8.1 Attribute Final Inspection Data Sheet for the P Chart

PRODUCT DESCRIPTION: MAGNETIC SENSOR
PART NUMBER: JHS041-11
OPERATION PROCESS: FINAL INSPECTION

A. FIRST NINE SAMPLE LOTS

Date	6/1/2009	6/1/2009	6/1/2009	6/2/2009	6/2/2009	6/2/2009	6/3/2009	6/3/2009	6/3/2009
Shift	1	2	3	1	2	3	1	2	3
Inspector	LBK	DRA	STY	LBK	DRA	STY	LBK	DRA	STY
Sample lot size, N	100	100	100	100	100	100	100	100	100
Sample lot no.	1	2	3	4	5	6	7	8	9
Total defective products, NP	9	6	11	14	13	6	14	10	6
Proportion of defective products to sample size = NP/N	0.09	0.06	0.11	0.14	0.13	0.06	0.14	0.10	0.06
Percentage of defective products to sample size = $100 \times (NP/N)$	9	6	11	14	13	6	14	10	6

B. SAMPLE LOTS 10 THROUGH 18

Date	6/4/2009	6/4/2009	6/4/2009	6/5/2009	6/5/2009	6/5/2009	6/6/2009	6/6/2009	6/6/2009
Shift	1	2	3	1	2	3	1	2	3
Inspector	LBK	DRA	STY	LBK	DRA	TAH	LBK	DRA	TAH
Sample lot size, N	100	100	100	50	50	50	50	50	50
Sample lot no.	10	11	12	13	14	15	16	17	18
Total defective products, NP	10	10	15	11	10	30	5	7	18
Proportion of defective products to sample size = NP/N	0.10	0.10	0.15	0.11	0.10	0.30	0.05	0.7	0.18
Percentage of defective products to sample size = $100 \times (NP/N)$	10	10	15	11	10	30	5	7	18

(Continued)

Table 8.1 (Continued) Attribute Final Inspection Data Sheet for the P Chart

PRODUCT DESCRIPTION: MAGNETIC SENSOR

PART NUMBER: JHS041-11

OPERATION PROCESS: FINAL INSPECTION

C. SAMPLE LOTS 19 THROUGH 25

Date	6/1/2009	6/1/2009	6/1/2009	6/2/2009	6/2/2009	6/2/2009	6/3/2009	6/3/2009	6/3/2009
Shift	1	2	3	1	2	3	1	2	3
Inspector	LBK	DRA	TAH	LBK	DRA	TAH	LBK		
Sample lot size, N	100	100	100	100	100	100	100		
Sample lot no.	19	20	21	22	23	24	25		
Total defective products, NP	11	12	17	11	9	14	9		
Proportion of defective products to sample size = NP/N	0.11	0.12	0.17	0.11	0.09	0.14	0.09		
Percentage of defective products to sample size = $100 \times (NP/N)$	11	12	17	11	9	14	9		

Then we have to calculate the process average proportion, $\bar{P}$, of nonconforming or defective products to sample size; for the initial P chart we use 25 sample lots.

$$\bar{P} = \frac{\sum\limits_{i=1}^{i=25} NP_i}{\sum\limits_{i=1}^{i=25} N} \tag{8.2}$$

Our average sample lot size for 25 lots is as follows:

$$\bar{N} = \frac{1}{25} \times \sum\limits_{i=1}^{i=25} N \tag{8.3}$$

By using the average proportion, $\bar{P}$, from Equation 8.2 and the average lot size, $\bar{N}$, from Equation 8.3, we can calculate the upper, UCL_P, and the lower, LCL_P, control limits for our P chart.

$$UCL_P = \bar{P} + 3 \times \sqrt{\frac{\bar{P} \times (1 - \bar{P})}{\bar{N}}} \tag{8.4}$$

and

$$LCL_P = \bar{P} - 3 \times \sqrt{\frac{\bar{P} \times (1 - \bar{P})}{\bar{N}}} \tag{8.5}$$

In some cases when $\bar{P}$ (PBAR) and $\bar{N}$ (NBAR) or both have small values, the lower control limit, LCL_P, calculated from Equation 8.5, can come out negative. In such cases you should use $LCL_P = 0$. Figure 8.1 shows when LCL_P goes negative for small values of $\bar{P}$ (PBAR) and $\bar{N}$ (NBAR) or both.

After obtaining the proportion of defective products from 25 sample lots we can plot the P chart for attribute data sheet as partially given in Table 8.1. Figure 8.2 shows the proportions of defective products for every sample lot, the average proportion of defective products for 25 sample lots, and the upper and lower control limits for the P chart for 25 sample lots. From sample lot 13 through sample lot 18, our sample sizes dropped from 100 products to 50 products due to a shortage of

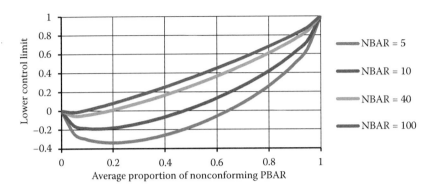

Figure 8.1 P chart lower control limit negative values at low average proportions and low average sample sizes. When LCL$_p$ is negative use LCL$_p$ = 0.

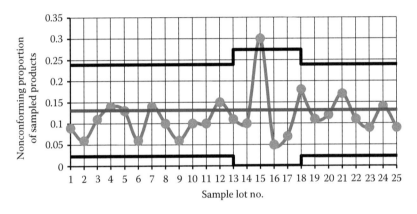

Figure 8.2 P chart for nonconforming proportion of 25 sampled products in the final inspection. (For 6 shifts—sample lot 13 through 18—upper and lower control limits are recalculated, since the sample lot size dropped from 100 to 50 during these shifts.)

a critical component from our vendor. Because sample sizes varied more than ±25% from the standard and agreed on sample size of 100, upper and lower control limits for sample sizes of 50—for 6 shifts— are recalculated.

The average proportion, $\bar{P}$, of defective products relative to sample size for attribute data in Table 8.1—for the initial 25 sample lots—is calculated using Equation 8.2 as follows:

$$\bar{P} = \frac{9+6+11+\cdots+15+11+10+30+5+7+18+11+\cdots+9+14+9}{100+100+100+\cdots+100+50+50+50+50+50+50+100+\cdots+100+100+100}$$

$$= 0.131$$

$$(8.6)$$

The average sample lot size for 25 lots of attribute data in Table 8.1 is calculated using Equation 8.3 as follows:

$$\bar{N} = \frac{1}{25} \times (100 + 100 + 100 + 100 + 100 + 100 + 100 + 100 + 100$$

$$+ 100 + 100 + 100 + 50 + 50 + 50 + 50 + 50 + 50 + 100$$

$$+ 100 + 100 + 100 + 100 + 100 + 100) = 88 \qquad (8.7)$$

The upper, UCL_P, and the lower, LCL_P, control limits for attribute data in Table 8.1—for the initial 25 sample lots—is calculated using Equations 8.4 and 8.5 as follows:

$$UCL_p = 0.131 + 3 \times \sqrt{\frac{0.131 \times (1 - 0.131)}{88}} = 0.239 \qquad (8.8)$$

and

$$LCL_p = 0.131 - 3 \times \sqrt{\frac{0.131 \times (1 - 0.131)}{88}} = 0.023 \qquad (8.9)$$

Now we have to calculate the upper, UCL_P, and the lower, LCL_P, control limits for attribute data from sample lot 13 through 18 in Table 8.1—for sample lot sizes of 50—again using Equations 8.4 and 8.5 as follows. The reason for this recalculation is the small sample lot size of 50 products. Our standard sample lot size in Table 8.1 is 100 products, and these six oddball sample lot sizes were more than 25% smaller than our standard sample size. For these six sample lots from 13 through 18 $\bar{P} = 0.131$ and $\bar{N} = 50$.

$$UCL_{\text{Sample lot 13 through 18}} = 0.131 + 3 \times \sqrt{\frac{0.131 \times (1 - 0.131)}{50}} = 0.274$$

$$(8.10)$$

and

$$LCL_{\text{Sample lot 13 through 18}} = 0.131 - 3 \times \sqrt{\frac{0.131 \times (1 - 0.131)}{50}} = -0.012$$

(8.11)

Since $LCL_{\text{Sample lot 13 through 18}}$ is negative, we should use zero as the lower control limit. The P chart for the nonconforming proportion of 25 sampled products in final inspection is shown in Figure 8.2.

As we can see from Figure 8.2, handling varying upper and lower control limits can be very cumbersome and confusing to operators. Therefore it is highly recommended to have constant or at most ±25% varying sample lot sizes for attribute P charts. We have the initial P chart for 25 sample lots in Figure 8.2. Now let us review this chart for statistical control.

Attribute control charts should be investigated similar to variable control charts for in-control or out-of-control conditions as explained in detail in Chapter 6. Before we can accept that our processes are in control, we have to determine causes for

- Points outside of control limits
- Eight points or more data runs in a succession on either side of the average
- Increasing or decreasing data trends successively in six intervals
- Cyclic nonrandom patterns
- Nonrandom patterns close to the centerline (more than two-thirds of points on the control chart)
- Data patterns where more than two data points in succession are close to upper or lower control limits

Once we determine causes for the anomalies above—out-of-control conditions—in our control charts, we have to find special causes that instigated them and stabilize our production process. Then we can generate a new set of control charts that will represent normal causes of variation in our processes. After we have clean and in statistical control charts without any anomalies, we can determine the process capability of our attribute data in final inspection.

In reviewing Figure 8.2, we see that one point—inspection results of sample lot number 15—is outside the upper control limit. We have to investigate to find and to eliminate special causes of variation that caused this out-of-control point. The quality control team investigates the out-of-control conditions immediately. The quality control team observes from data sheets in Table 8.1 that this out-of-control data point corresponds to the sample lot taken during the third shift by a new inspector with the initials TAH. The team agrees that this out-of-control condition is not related to a product quality issue, but stems from the new third shift operator. The corrective action decided by the quality control team is to retrain this new inspector. After extensive retraining of the third shift inspector, defective proportions of sample lots settle down to in statistical control conditions. The quality control team gets this sample lot 15 reinspected. Then they recalculate $\bar{P}$, $\bar{N}$, and upper (UCL_p) and lower (LCL_p) control limits to continue to monitor final inspection results. The updated and in statistical control P chart is presented in Figure 8.3.

Now that our P chart is in statistical control, we can investigate results from the aforementioned P chart for process capability. Process capability interpretation is similar to variable control charts, which are discussed in detail in Chapter 10. From Figure 8.3 we see that on the average 12.0% of magnetic sensors are rejected due to attribute defects, namely $\bar{P} = 12.0\%$. The standard deviation or spread of defective magnetic sensors during the final inspection is 3.5%, namely

$$\sigma = \frac{(UCL_p - \bar{P})}{3} \text{ or } \sigma = \frac{(22.5\% - 12.0\%)}{3} = 3.5\%.$$

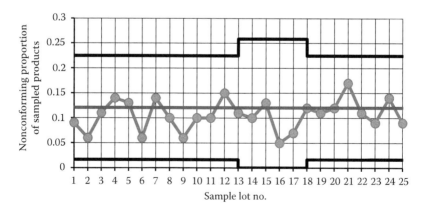

Figure 8.3 In statistical control P chart for nonconforming proportion of sampled products in final inspection. (Out-of-control sample lot 15 is reinspected.)

Next let us put together another attribute chart. This attribute chart is an NP chart for nonconforming or defective products in an inspection lot. NP charts are identical to P charts except that the actual number of nonconforming or defective products—instead of proportions—is plotted and analyzed. For NP chart data collection, sample lot sizes should stay a constant. Data for an NP chart for defective products from weld inspection are provided in Table 8.2.

Let us now plot the NP chart for defective products from the weld inspection provided in Table 8.2 for 25 sample lots of equal size, namely $N = 200$. The average number of defective products per sample lot, $\overline{NP}$, is

$$\overline{NP} = \left(\frac{1}{25}\right) \times \sum_{i=1}^{25} NP_i = \frac{6+3+8+\cdots 9+8+9}{25} = 8.80 \quad (8.12)$$

The upper and lower control limits for the NP chart are calculated as follows:

$$\text{UCL}_{NP} = \overline{NP} + 3 \times \sqrt{\overline{NP} \times (1 - \bar{P})}$$

$$= 8.80 + 3\sqrt{8.80 \times \left(1 - \frac{8.80}{200}\right)} = 17.50 \quad (8.13)$$

and

$$\text{LCL}_{NP} = \overline{NP} - 3 \times \sqrt{\overline{NP} \times (1 - \bar{P})}$$

$$= 8.80 - 3\sqrt{8.80 \times \left(1 - \frac{8.80}{200}\right)} = 0.099 \quad (8.14)$$

Now we can plot the NP chart shown in Figure 8.4 for defective products from the weld inspection using $\overline{NP}$, UCL_{NP}, and LCL_{NP} values calculated previously.

First we have to analyze the NP chart from the preceding text for statistical control. We use the same criteria as before in P charts and in variable control charts to establish if we have any out-of-control

Table 8.2 Attribute Weld Inspection Data Sheet for the NP Chart

PRODUCT DESCRIPTION: BASE STRUCTURE
PART NUMBER: JHS041

OPERATION PROCESS: WELD INSPECTION

A. FIRST 9 SAMPLE LOTS

Date	7/11/2011	7/12/2011	7/13/2011	7/14/2011	7/15/2011	7/18/2011	7/19/2011	7/20/2011	7/21/2011
Inspector	Adam	Adam	Adam	Adam	Adam	Adam	Adam	Adam	Adam
Sample lot size, N	200	200	200	200	200	200	200	200	200
Sample lot no.	1	2	3	4	5	6	7	8	9
No. of defective products, NP	6	3	8	11	13	4	9	10	6

B. SAMPLE LOTS 10 THROUGH 18

Date	7/22/2011	7/25/2011	7/26/2011	7/27/2011	7/28/2011	7/29/2011	8/1/2011	8/2/2011	8/3/2011
Inspector	Adam	Adam	Adam	Adam	Adam	Adam	Adam	Adam	Adam
Sample lot size, N	200	200	200	200	200	200	200	200	200
Sample lot no.	10	11	12	13	14	15	16	17	18
No. of defective products, NP	5	10	12	11	10	6	5	7	15

C. SAMPLE LOTS 19 THROUGH 25

Date	8/4/2011	8/5/2011	8/8/2011	8/9/2011	8/10/2011	8/11/2011	8/12/2011
Inspector	Adam	Adam	Adam	Adam	Adam	Adam	Adam
Sample lot size, N	200	200	200	200	200	200	200
Sample lot no.	19	20	21	22	23	24	25
No. of defective products, NP	11	12	9	11	9	8	9

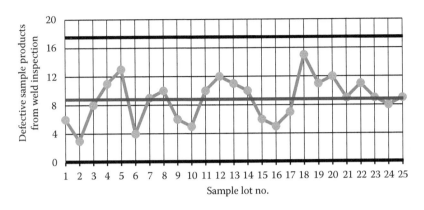

Figure 8.4 NP chart for defective sample products from the weld inspection (sample size = 200).

conditions. These out-of-control conditions can have special causes such as variations in our processes, errors in our plotting, changes in our measurement system, and so forth, while taking the first 25 data points in Figure 8.2. Six criteria for out-of-control conditions are

- Points outside of control limits
- Eight points or more data runs in a succession on either side of the average
- Increasing or decreasing data trends successively in six intervals
- Cyclic nonrandom patterns
- Nonrandom patterns close to the centerline (more than two-thirds of points on the control chart)
- Data patterns where more than two data points in succession are close to the upper or lower control limits

The NP chart in Figure 8.4 seems to be in statistical control. So we can continue to gather our welding inspection data and keep plotting onto the NP chart by using the initial values for NP, UCL_{NP}, and LCL_{NP} calculated previously.

We can also perform a process capability analysis similar to the one for the P charts, since the control chart in Figure 8.4 is in statistical control. From Figure 8.4 we see that on average 8.8 of base structures are rejected due to weld defects, namely $\bar{P} = 8.8$. The standard deviation or spread of defective base structures during weld inspection is 2.9, namely $\sigma = \dfrac{(UCL_p - \bar{P})}{3}$ or $\sigma = \dfrac{(17.5 - 8.8)}{3} = 2.9.$

If the initial 25 points in Figure 8.4 showed any out of statistical control condition, then we had to determine special causes that generated the out of statistical control condition. We had to correct for special causes—stabilize our production processes, correct for control charting errors, stabilize our measurement system, and so forth—before we can start to generate a new NP chart.

Another type of attribute control chart is the C chart. The C chart is used for a number of nonconformities or defects in a constant sized sample lot. It does not reflect the number of defective products as in P charts and in NP charts. The C chart focuses on quantity of nonconformities or defects found in inspected unit(s). Sample lots should have a constant size. An example is the number of visual defects on a wafer at final inspection. Another example is the number of paint defects on a finished vehicle. One more example is the number of patients with high blood pressure for every hundred tested.

The average number of nonconformities or defects in inspected samples is

$$\bar{C} = \left(\frac{1}{N} \right) \times \sum_{i=1}^{N} C_i \tag{8.15}$$

$C_1, C_2, C_3, \ldots, C_N$ are nonconformities or defects in each sample and N is the number of samples. Upper and lower control limits for a C chart are calculated as follows:

$$\text{UCL}_C = \bar{C} + 3 \times \sqrt{\bar{C}} \tag{8.16}$$

and

$$\text{LCL}_C = \bar{C} - 3 \times \sqrt{\bar{C}} \tag{8.17}$$

The number of contaminated regions on a completed sample 6-inch wafer is recorded during every shift to control cleanliness in our wafer processes. The C chart in Figure 8.5 shows the inspection results for 30 shifts. First you need to check the C chart in Figure 8.5 for statistical control. We see that there are six increasing intervals between sample wafer 18 and sample wafer 24. The C chart is out of control

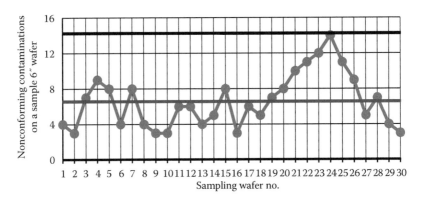

Figure 8.5 C chart for nonconforming contaminations on a sample 6-inch wafer.

during this period. We have to find special cause(s) in our processes that instigated this increasing trend. The control chart response team for this C chart puts together a plan to attack this out-of-control trend. They found the culprit(s), corrected for it (them), and the control chart data started to go back to normal behavior. After the wafer processes settle down, we should eliminate data points from sample wafer 18 to sample wafer 24, inspect six more wafers for contamination, one per shift; recalculate $\bar{C}$, UCL_C, and LCL_C; and continue monitoring contaminants on our wafers using this C chart.

We can also perform a process capability analysis similar to the one for the P charts after we get the C chart in Figure 8.5 in statistical control.

There is another attribute chart, called the U chart, which I discuss next. Like C charts, U charts control the number of nonconformities or defects per sample unit. However, in U charts, sample sizes can vary. U charts control defects similar to C charts, except that sample units do not have a constant size.

We plot the number of nonconformities or defects per unit, U, in every sample lot.

$$U = \frac{NC}{N} \tag{8.18}$$

where NC is the total number of nonconformities or defects in N units in a sample lot. Then the average number of nonconformities or defects are inspected—j—sample lots for our control charting is

$$\bar{U} = \frac{\sum_{i=1}^{j} NC_i}{\sum_{i=1}^{j} N_i} \tag{8.19}$$

$NC_1, NC_2, NC_3, \ldots, NC_j$ are nonconformities or defects in each sample lot and $N_1, N_2, N_3, \ldots, N_j$ are number of units in each sample lot. Upper and lower control limits for a U chart is calculated as follows:

$$\text{UCL}_U = \bar{U} + 3 \times \sqrt{\frac{\bar{U}}{\bar{N}}} \tag{8.20}$$

and

$$\text{LCL}_U = \bar{U} - 3 \times \sqrt{\frac{\bar{U}}{\bar{N}}} \tag{8.21}$$

where $\bar{N} = \left(\frac{1}{j}\right) \times \sum_{i=1}^{j} N$ is the average sample lot size.

A U chart for the number of chips and scratches found on each finished chess board is shown for 30 sample lots with different lot sizes in Figure 8.6. First you need to check the U chart in Figure 8.6 for statistical control using the same guidelines as for P charts given at the beginning of this chapter. We see that there are 11 points in a row

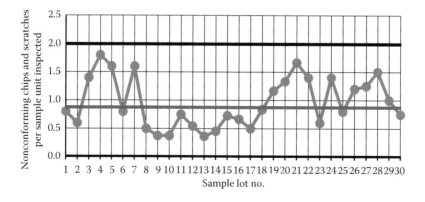

Figure 8.6 U chart for nonconforming chips and scratches on chess boards per sample unit inspected.

lower than and close to the average between sample lot 8 and sample lot 18. The U chart is out of control during this period in a good way. We have to find special cause(s) in our processes that lowered defect quantities per unit during this period. The control chart response team for this U chart puts together a plan to attack this out-of-control trend. They found the good culprit(s), corrected for it (them), and the control chart data started to go back to normal behavior. Hopefully the quantity of chips and scratches per unit we are looking for in each sample lot should go down. After improved chess board production processes settle down, we should eliminate data points from sample wafer 8 to sample wafer 18; inspect 11 more new sample lots; recalculate $\bar{U}$, UCL_U, and LCL_U; and continue monitoring chips and scratches on our finished chess boards using this U chart.

We can also perform a process capability analysis similar to the one for the P charts after we get the U chart in Figure 8.6 in statistical control.

9

CORRELATION BETWEEN TWO GAGES

When we are dealing with two or more gages measuring the same parameter, we have to determine if the measurements made by these gages correlate accurately with each other. These correlation exercises can be performed on gages within your company's divisions, between your company's and your customer's divisions, or between your company's and your subcontractor's divisions.

Before taking any data for gage correlation, we have to make sure that gages are calibrated to well established standards (see Chapter 1), measurement procedures are well defined and released officially in document control, and the monitor control chart for each gage is in control. Data taken for correlation purposes should cover a range beyond the critical parameter's specifications.

For example, if we are measuring a gap length of 30 ± 5 microinches with our gages, the correlation data range should cover 20 to 40 microinches. A common mistake that is made in correlating gages is to assume that the correlation being performed is applicable beyond the range of the data in hand.

Here are some examples of correlation data between two gages:

- Figure 9.1 represents a linear correlation with a positive slope and with an accurate relationship between two gages.
- Figure 9.2 also represents a linear correlation with a positive slope, but with an inaccurate relationship between two gages.
- Figure 9.3 represents a nonlinear correlation with a positive slope and with an accurate relationship between two gages.
- Figure 9.4 represents a linear correlation with a negative slope and with an inaccurate relationship between two gages.
- Figure 9.5 represents almost no correlation between two gages.

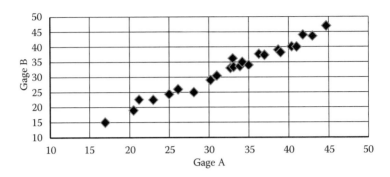

Figure 9.1 Linear correlation with positive slope between gages A and B and with an accurate relationship.

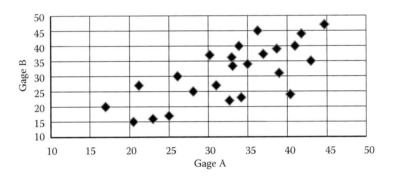

Figure 9.2 Linear correlation with positive slope between gages A and B and with an inaccurate relationship.

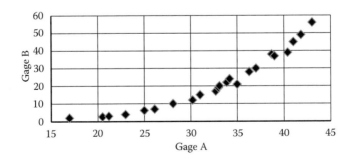

Figure 9.3 Nonlinear correlation with positive slope between gages A and B and with an accurate relationship.

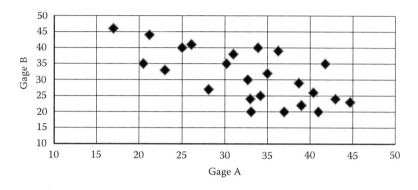

Figure 9.4   Linear correlation with negative slope between gages A and B and with an inaccurate relationship.

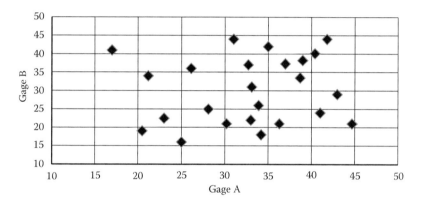

Figure 9.5   No correlation between gages A and B.

In Figure 9.1, data points seem to form a linear correlation with a positive slope. Also, data points are not spreading out much around a linear line. Later in this chapter, we will determine the best fit linear line or line of regression for these data points.

In Figure 9.2, data points between gages A and B seem to form a correlation with a positive slope. However, in this example data points are spread out. It is hard to represent an accurate correlation between two gages.

In Figure 9.3, data points seem to form a nonlinear correlation with a positive slope. Also, data points are not spreading out much around a nonlinear line.

In Figure 9.4, data points between gages A and B seem to form a correlation with a negative slope. However, also in this example data points are spread out. Again, it is hard to represent an accurate correlation between two gages.

For correlation data in Figure 9.5, there seems to be no correlation. Data are spread out all over the graph. A reading with gage A can be any arbitrary reading with gage B. We cannot see any correlation between the two gages.

Next let us determine the best fit linear line for data points obtained to correlate our two gages. We will achieve this goal by minimizing squares of distances, namely d_i^2's, between all data points and the best fit curve as shown in Figure 9.6. This is called the method of least squares in the statistics literature.

A best fit linear line can be represented by

$$\text{Gage B} = m + n \times \text{Gage A} \tag{9.1}$$

where m is the intercept of the best fit linear line and n is the slope of the best fit linear line. Squares of distances, namely the sum of all d_i^2's (see Figure 9.6), between all data points and the best fit curve can be written as

$$d^2 = \sum_{i=1}^{j} d_i^2 = \sum_{i=1}^{j} [\text{Gage B}_i - (m + n \times \text{Gage A}_i)]^2 \tag{9.2}$$

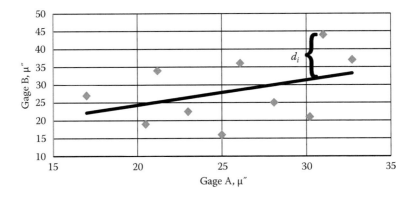

Figure 9.6   Best fit linear line to correlation data.

To determine the minimum squares of distances, namely the minimum of the sum of d_i^2's, we have to take the partial derivative of d^2 with respect to m and also with respect to n and then we have to set these resulting Equations 9.3 and 9.4 to zero to determine the values of m and n for the best fit linear line.

$$\frac{\partial d^2}{\partial m} = -2\sum_{i=1}^{j}[\text{Gage B}_i - (m + n \times \text{Gage A}_i)] = 0 \qquad (9.3)$$

$$\frac{\partial d^2}{\partial n} = -2\sum_{i=1}^{j}[\text{Gage B}_i - (m + n \times \text{Gage A}_i)] \times \text{Gage A}_i = 0 \quad (9.4)$$

Solving for n from Equations 9.3 and 9.4 and then using Equation 9.1 provides the following results that give us the best fit linear line.

$$n = \frac{\sum_{i=1}^{j}\text{Gage A}_i \times \text{Gage B}_i - j \times \text{Gage A}_{\text{AVE}} \times \text{Gage B}_{\text{AVE}}}{\sum_{i=1}^{j}\text{Gage A}_i \times \text{Gage A}_i - j \times \text{Gage A}_{\text{AVE}} \times \text{Gage A}_{\text{AVE}}}$$
$$(9.5)$$

where

$$\text{Gage A}_{\text{AVE}} = \left(\frac{1}{j}\right) \times \sum_{i=1}^{j}\text{Gage A}_i \text{ and Gage B}_{\text{AVE}} = \left(\frac{1}{j}\right) \times \sum_{i=1}^{j}\text{Gage B}_i$$

$$m = \text{Gage B}_{\text{AVE}} - n \times \text{Gage A}_{\text{AVE}} \qquad (9.6)$$

From Equations 9.5 and 9.6, we can determine the best fit linear line parameters for the correlation data in Figure 9.6 by creating a simple table, shown in Table 9.1, for the number of data points $j = 10$.

Table 9.1   Sums and Averages of Correlation Data Needed to Calculate Best Fit Linear Line Parameters Using Equations 9.5 and 9.6

GAGE A	GAGE B	GAGE A × GAGE B	GAGE A × GAGE A
17.0	27.0	459.0	289.00
20.5	19.0	389.5	420.25
21.2	34.0	720.8	449.44
23.0	22.5	517.5	529.00
25.0	16.0	400.0	625.00
26.1	36.0	939.6	681.21
28.1	25.0	702.5	789.61
30.2	21.0	634.2	912.04
31.0	44.0	1364	961.00
32.7	37.0	1209.9	1069.29
Gage $A_{AVE} = 25.48$	Gage $B_{AVE} = 28.15$	Sum (Gage A × Gage B) = 7337.0	Sum (Gage A × Gage A) = 6725.84

Equation 9.5 gives us the slope of the best fit linear line for data in Table 9.1 as follows:

$$n = \frac{\displaystyle\sum_{i=1}^{j} \text{Gage A}_i \times \text{Gage B}_i - j \times \text{Gage A}_{AVE} \times \text{Gage B}_{AVE}}{\displaystyle\sum_{i=1}^{j} \text{Gage A}_i \times \text{Gage A}_i - j \times \text{Gage A}_{AVE} \times \text{Gage A}_{AVE}}$$

$$= \frac{7337.0 - 10 \times 25.48 \times 28.15}{6725.84 - 10 \times 25.48 \times 25.48} = 0.7039 \qquad (9.7)$$

Equation 9.6 gives us the intercept of the best fit linear line for data in Table 9.1 as follows:

$$m = \text{Gage B}_{AVE} - n \times \text{Gage A}_{AVE} = 28.15 - 0.7039 \times 25.48 = 10.215 \qquad (9.8)$$

or

$$\text{Gage B}_{\text{best fit linear line}} = 10.215 + 0.7039 \times \text{Gage A} \qquad (9.9)$$

We can also obtain the best fit linear line, Equation 9.9, by using MS Excel®. After plotting the correlation data using the XY (Scatter) chart in MS Excel, we can highlight the correlation data and choose "Add Trendline" to the chart. Then we highlight "Linear" and "Display Equation on Chart" to get Equation 9.9.

Next let us look at the spread of correlation data points around the best fit linear line. This spread or dispersion of correlation data around the best fit linear line will be an indication of how good a correlation we have between gages A and B. In Figure 9.6 there is a substantial amount of spread between the observed correlation data points and the best fit linear line. One of the ways to evaluate the goodness of correlation between two gages is to calculate the standard error of estimate between the actual correlation data and the best fit linear line as follows:

$$S_{\text{standard error of estimate}} = \sqrt{\frac{\sum_{i=1}^{j}(\text{Gage B} - \text{Gage B}_{\text{best fit linear line}})^2}{j-2}} \tag{9.10}$$

The standard error of estimate for the correlation exercise in Figure 9.6 or Table 9.1 is 8.912 units. Another way to look at this spread is that 99.7% of gage B correlation data can be within ±26.735 units (3 × 8.912) of best fit linear line for a certain gage A value. So it is hard to predict what gage B is going to be when a gage A data point is given.

The goodness of linear correlation between gage A and gage B can also be described by the coefficient of determination. The strength of the correlation between gages A and B—coefficient of determination—can be defined as follows:

$$R^2 = 1 - \frac{\sum_{i=1}^{j}(\text{Gage B} - \text{Gage B}_{\text{best fit linear line}})^2}{\sum_{i=1}^{j}(\text{Gage B} - \text{Gage B}_{\text{AVE}})^2} \tag{9.11}$$

The coefficient of determination for the correlation exercise in Figure 9.6 or Table 9.1 is 0.1541. We can also obtain the coefficient of determination, Equation 9.11, by using MS Excel. After plotting correlation data using the XY (Scatter) chart in MS Excel, we can highlight the correlation data on the chart and choose "Add Trendline" to the chart. Then we highlight "Linear" and "Display R-squared Value on Chart" to get Equation 9.11.

When all correlation data points lie on the best fit linear line, there is a perfect correlation between two gages, namely $R^2 = 1$. However, when $R^2 \to 0$, there is no correlation between two gages. The coefficient of determination for the correlation exercise in Figure 9.6 or Table 9.1 is 0.1541, which means that there is a very weak correlation between two gages. For a strong linear relationship between two gages, $R^2 > 0.8$.

Let us revisit Figures 9.1 through 9.5 and determine the best fit linear line for each type of gage correlation—called trend line in MS Excel—and calculate the coefficient of determination—called R-squared value in MS Excel—for every case. Figure 9.7 represents the correlation data given in Figure 9.1.

Figure 9.7 shows a strong linear correlation betweeen two gages with a high R^2 value of 0.9763. However, the opposite is true for the correlation data given in Figure 9.2. The best fit linear line has a weak R^2 value of 0.5287, as shown in Figure 9.8.

In Figure 9.3, data points seem to form a nonlinear correlation with a positive slope. Also, data points are not spreading out much around a nonlinear line. We can reduce this nonlinear relationship to a linear one by using logarithmic coordinates. Then we can still determine the goodness of the linear correlation between gage A and

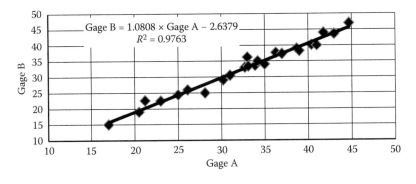

Figure 9.7   Data in Figure 9.1 show positive and strong linear correlation between gages A and B.

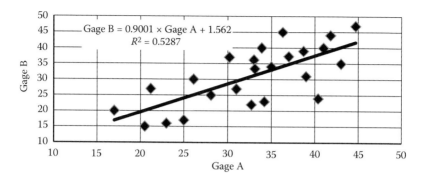

Figure 9.8 Data in Figure 9.2 show positive and weak linear correlation between gages A and B.

gage B by the coefficient of determination. The nonlinear power relationship is as follows:

$$\text{Gage B} = e \times (\text{Gage A})^f \tag{9.12}$$

or in logarithmic coordinates

$$\text{Log}_{10}(\text{Gage B}) = \text{Log}_{10}(e) + f \times \text{Log}_{10}(\text{Gage A}) \tag{9.13}$$

When we compare Equation 9.13 to the linear Equation 9.1, we observe that the intercept $m = \text{Log}_{10}(e)$ and the slope $n = f$. The correlation data in Figue 9.3 are shown in logarithmic coordinates in Figure 9.9. The correlation data in Figure 9.3 show a positive and strong nonlinear correlation between gages A and B as depicted in Figure 9.10, where $e = 10^{-4.6591} = 0.000022$ and $f = 3.9093$. The correlation data in Figure 9.4 represent a weak relationship between gages A and B with a negative slope, as shown in Figure 9.11. The correlation

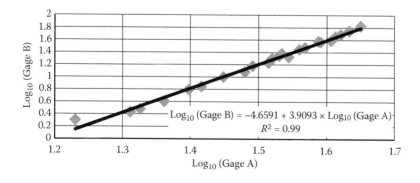

Figure 9.9 Correlation data in Figure 9.3 are linearized by using logarithmic coordinates.

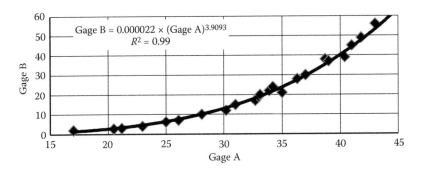

Figure 9.10 Correlation data in Figure 9.3 show positive and strong nonlinear correlation between gages A and B, namely $R^2 = 0.99$.

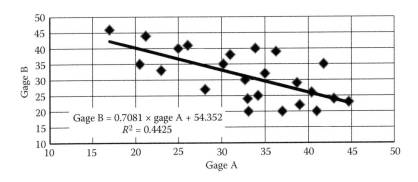

Figure 9.11 Linear correlation with a negative slope between gages A and B with a weak coefficient of determination of 0.4425.

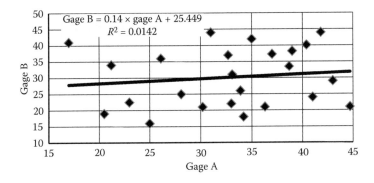

Figure 9.12 No correlation between gages A and B in Figure 9.5 with an R^2 value of almost zero.

data in Figure 9.5 are all scattered and show no correlation between gages A and B, as verified in Figure 9.12.

There is another important factor in correlating two gages. This factor is the "significance" of correlation between the two gages. We have to take enough correlation data points—calling the number of data points j—so that $R\sqrt{j}$ should be greater than 3 for a significant correlation between two gages. Also, correlation data points should cover a range beyond our specification's tolerances. For example, in Figure 9.4 we have a strong correlation between two gages: $R^2 = 0.9763$ with $j = 24$. The significance of the correlation value for this case becomes $R\sqrt{j} = 4.84 > 3$.

In another example, the coefficient of determination for the correlation exercise in Figure 9.6 or Table 9.1 is 0.1541 and therefore the correlation is very weak. At the same time the correlation between two gages is insignificant because of insufficient data, namely $j = 10$ and $R\sqrt{j} = 1.24 < 3$.

10

PROCESS CAPABILITY

Once our variable control charts are in statistical control for a certain variable as described in Chapters 6, 7, and 8, we can determine the process capability of our product for that variable. Most variable characteristics of a product behave like a normal distribution. There might be some variable characteristics of a product that might show skewed, bimodal, or some other nonnormal distribution behavior. However, nonnormal distribution behaviors for a variable are very rare in high-volume production.

Here we first show as an example that the population of our sensor that we analyzed in Chapter 6 shows a normal distribution behavior for its resistance. Since the mean of sample averages, $\bar{X}$, is the same as the mean for the whole product population and since the spread of product population can be determined from the mean of sample ranges, $\bar{R}$—for process data in Table 6.1, $\bar{\bar{X}} = 24.941$ Ohms and $\bar{R} = 0.546$ Ohms—we can compare the frequency distribution of resistance data for the first 125 samples to the population normal distribution.

The process standard deviation can be estimated from the mean of sample ranges as follows:

$$\sigma = \frac{\bar{R}}{d_2} \tag{10.1}$$

where d_2 depends on sample size and it can be found in A. J. Duncan's *Quality Control and Industrial Statistics* (1974), p. 968, Table M. For a sample size of 5, $d_2 = 2.326$. The constants d_2 for different sample sizes are also given in Table 10.1.

We can first organize sensor resistance data from Table 6.1 into bins—binning 125 data points—to obtain the frequency distribution. Frequency distribution data and the normal distribution values (see

Table 10.1 Constants d_2 for Process Standard Deviation Estimation from Mean of Sample Ranges

SAMPLE SIZE	d_2
2	1.128
3	1.693
4	2.059
5	2.326
6	2.534
7	2.704
8	2.847
9	2.970
10	3.078

Note: Factors A, A_2, B_3, B_4, d_2, $\frac{1}{d_2}$, d_3, and D_1–D_4 are reproduced with permission from Table B2 of the ASTM Manual on Quality Control of Materials, January 1951, p. 115. Factors A_3, B_5, B_6, and c_4 are reproduced with permission from American Society for Quality (ASQC) Standard A1, Table 1.

For control chart factors, most text books refer to ASTM Manual on Quality Control of Materials, January 1951, p. 115, Table B2.

Chapter 5) are presented in Table 10.2. Normal distribution values are obtained using $\overline{\overline{X}} = 24.941$ Ohms and $\sigma = \dfrac{0.546}{2.326} = 0.235$ Ohms.

The easiest way to visualize if our variable data are behaving as a normal distribution is to graph them together as presented in Figure 10.1.

The frequency distribution for the sensor resistance population shows pretty much a normally distributed behavior. The frequency distribution is almost symmetrical and shows little positive skewness. For practical purposes we can definitely use the normal distribution to estimate our process capability for our sensor resistance.

For a sensor resistance specification of 25 ± 0.5 Ohms, for a population average resistance of $\overline{\overline{X}} = 24.941$ Ohms, and for a process standard deviation of $\sigma = \dfrac{0.546}{2.326} = 0.235$ Ohms, we can see in Figures 10.2 through 10.4 that our process is not capable. We can calculate the percentage and the amount of our sensors that will be below 24.5 Ohms by using cumulative probabilities of normal distribution tables in statistics books or by using the NORMDIST statistical formula in MS Excel®. A good reference table for cumulative

Table 10.2   Frequency and Normal Distribution Values for Sensor Resistance Data from Table 6.1

FREQUENCY DISTRIBUTION BINS FOR SENSOR RESISTANCE, OHMS	NO. OF DATA POINTS IN THE BIN	SENSOR RESISTANCES FOR NORMAL DISTRIBUTION CALCULATIONS, OHMS	NORMAL DISTRIBUTION VALUES
24.3 < Resistance < 24.4	0	24.3	0.0405
24.4 < Resistance < 24.5	0	24.4	0.1188
24.5 < Resistance < 24.6	0	24.5	0.2900
24.6 < Resistance < 24.7	6	24.6	0.5905
24.7 < Resistance < 24.8	7	24.7	1.0024
24.8 < Resistance < 24.9	17	24.8	1.4189
24.9 < Resistance < 25.0	29	24.9	1.6747
25.0 < Resistance < 25.1	23	25.0	1.6482
25.1 < Resistance < 25.2	14	25.1	1.3526
25.2 < Resistance < 25.3	11	25.2	0.9254
25.3 < Resistance < 25.4	9	25.3	0.5279
25.4 < Resistance < 25.5	5	25.4	0.2511
25.5 < Resistance < 25.6	4	25.5	0.0996
25.6 < Resistance < 25.7	0	25.6	0.0329
25.7 < Resistance < 25.8	0	25.7	0.0091
25.8 < Resistance < 25.9	0	25.8	0.0021

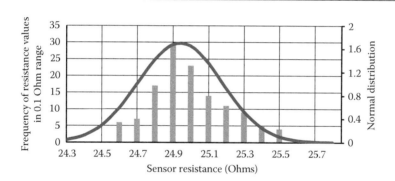

Figure 10.1   Frequency and normal distributions for initial sensor resistance data.

probabilities of normal distribution is A. J. Duncan's *Quality Control and Industrial Statistics* (1974), p. 945, Table A2. Cumulative probabilities of the normal distribution in statistics books are provided as a function of a nondimensional variable Z where $Z = \dfrac{(X - \bar{\bar{X}})}{\sigma}$. For our case, the nondimensional lower specification point is

$$Z = \frac{(X - \bar{\bar{X}})}{\sigma} = \frac{(24.5 - 24.941)}{0.235} = -1.877 \qquad (10.2)$$

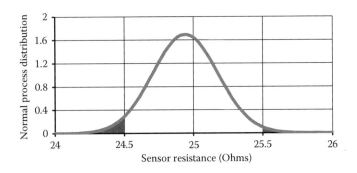

Figure 10.2 Sensor resistance process capability—red regions represent reject parts (left red region is for rejected parts below 24.5 Ohms and right red region is for rejected parts above 25.5 Ohms).

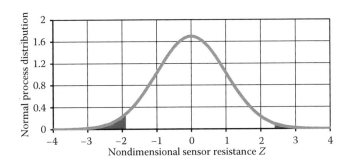

Figure 10.3 Sensor resistance process capability as a function of nondimensional sensor resistance Z. Red regions represent reject parts (left red region is for rejected parts below $Z = -1.877$ and right red region is for rejected parts above $Z = +2.379$).

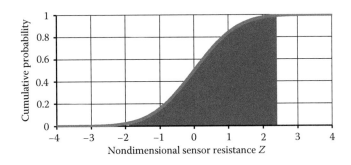

Figure 10.4 Cumulative probability of normal distribution for sensor resistance. Good products are between $Z = -1.877$ and $Z = +2.379$ values.

From A. J. Duncan's *Quality Control and Industrial Statistics* (1974), p. 945, Table A2, we can interpolate to get the cumulative probability for $Z = +1.877$ to be 0.9697. Since the normal distribution function is symmetrical, the cumulative probability from $Z \rightarrow -\infty$ to $Z = -1.877$ is the same as the cumulative probability from $Z \rightarrow +1.877$ to $Z \rightarrow +\infty$. Therefore the area under the normal distribution function (cumulative probability) from $Z \rightarrow -\infty$ to $Z = -1.877$ is $(1 - 0.9697)$ or 0.0303. Therefore 3.03% of our sensors will be below 24.5 Ohms by using this table.

By using MS Excel's NORMDIST function, we can get a more accurate answer easily by finding the area under the normal distribution function (cumulative probability) from $X \rightarrow -\infty$ to $X = 24.5$ Ohms.

$$\text{NORMDIST}(24.5, 24.941, 0.235, \text{TRUE}) = 0.031 \text{ or } 3.1\% \quad (10.3)$$

or

$$\begin{array}{c} \text{No. of sensors out of a million below the} \\ \text{24.5 Ohms specification is } 0.031 \times 10^6 = 30,580 \end{array} \quad (10.4)$$

We can also calculate the percentage and the amount of our sensors that will be above 25.5 Ohms. For our case, the nondimensional upper specification point is

$$Z = \frac{(X - \bar{\bar{X}})}{\sigma} = \frac{(25.5 - 24.941)}{0.235} = +2.379 \quad (10.5)$$

From A. J. Duncan's *Quality Control and Industrial Statistics* (1974), p. 945, Table A2, we can interpolate to get the cumulative probability for $Z = +2.379$ to be 0.9913. Therefore the area under the normal distribution function (cumulative probability) from $Z \rightarrow +2.379$ to $Z \rightarrow +\infty$ is $(1 - 0.9913)$ or 0.0087. Therefore 0.87% of our sensors will be above 25.5 Ohms by using this table.

By using MS Excel's NORMDIST function, we can get the answer more accurately and more easily by finding the area under the normal distribution function (cumulative probability) from 25.5 Ohms to $+\infty$ and subtracting it from unity, which is the area under the normal distribution function.

$$1 - \text{NORMDIST}(25.5, 24.941, 0.235, \text{TRUE})$$
$$= 0.0086 \text{ or } 0.86\% \qquad (10.6)$$

or

No. of sensors out of a million above the
25.5 Ohms specification = $0.0086 \times 10^6 = 8587$ (10.7)

The reject regions are shown graphically in Figures 10.2 and 10.3. The cumulative probability of good products is shown in Figure 10.4.

The aforementioned process capability analysis and Figures 10.2 through 10.4 show that our sensor processes are not capable of achieving our customer's resistance specification. We have to improve our processes to tighten spread of resistance in our sensors. After we perform several design-of-experiments to tighten our resistance spread in our sensors, we have to stabilize our newly improved processes. Then we can start to create new control charts for sensor resistance. We have to take the initial 25 sample groups in appropriate time intervals for our control charts. After the 25th sample group, we have to calculate upper and lower control limits for the $\bar{X}$ and R charts. We have to verify that our new processes are in control. After completing this cycle, we can reassess our process capability for our sensor resistance.

Our customer can request from us a 6σ capability for our sensor resistance for ship-to-stock qualification. This means that we have to reduce our processes spread by about threefold, namely to $\dfrac{(25.0 - 24.5)}{6} = 0.0833$ Ohms by $\dfrac{0.235}{0.0833} = 2.82$-fold. In 6σ capability, the population average is allowed to shift by $\pm 1.5\sigma$ about the mean. In this case, our sensor's population average can shift within $25 \pm 1.5 \times 0.0833$ or from 24.875 Ohms to 25.125 Ohms. Then we have 4.5σ left for our process capability. In one of the worst case scenarios, if our population average shifted lower to 24.875 Ohms, the number of sensors below 24.5 Ohms will be

$$\text{NORMDIST}(24.5, 24.875, 0.0833, \text{TRUE})$$
$$= 0.000003369 \text{ or } 0.0003369\% \qquad (10.8)$$

or

No. of sensors out of a million below the
24.5 Ohms specification = $3.369 \times 10^{-6} = 3.4$ (10.9)

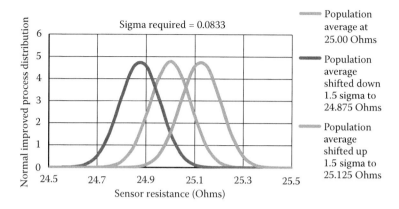

Figure 10.5   Improved process capability to 6σ with allowed ± 1.5σ population average shifts.

The same result will be obtained if our population average shifted higher to 25.125 Ohms, and the number of sensors above 25.5 Ohms will be 3.4 out of a million sensors. 6σ capable normal distributions are shown in Figure 10.5 with allowable population averages at nominal and extreme cases.

In process capability discussions, some people use the process capability index, C_{pk}, which is defined as follows:

$$C_{pk} = \frac{Z_{min}}{3}$$

$$(10.10)$$

where Z_{min} is the minimum of $\dfrac{(\text{Upper Specification Limit} - \bar{\bar{X}})}{\sigma}$ or

$-\dfrac{(\text{Lower Specification Limit} - \bar{\bar{X}})}{\sigma}$.

If $Z_{min} = 3$, we have a process capability index of 1 or a process capability of $\bar{\bar{X}} \pm 3\sigma$. If $Z_{min} = 4.5$, we have a process capability index of 1.5 or a process capability of $\bar{\bar{X}} \pm 4.5\sigma$, which is shown in the preceding text as the 6σ process capability requirement.

In some cases process capability can be used to estimate the quantity of parts that are to be produced for a given customer's specification. This type of usage for process capability can be explained best by an example.

Suppose we are producing precision steel balls. Our production processes are in control. The product's diameter shows a normal distribution shape. The population average for the diameter of this particular steel ball is

$$\bar{\bar{X}} = 20.00010 \, \text{mm} \tag{10.11}$$

and the standard deviation for the diameter of this particular steel ball is

$$\sigma = 0.00030 \, \text{mm} \tag{10.12}$$

Our customer wants 10,000 of these precision steel balls within his specification, namely 20.00000 ± 0.00060 mm. We compare our customer's specification to our process capability (see Figure 10.6), and we realize that we have to do 100% sorting for ball diameters under our automated laser beam gages to satisfy our customer's requirements.

The red regions represent ball diameter rejects from our process. The percentage of ball diameters below 19.99940 mm or $Z = -2.33$ can be calculated using MS Excel as follows:

$$\text{NORMALDIST}(19.99940, 20.00010, 0.00030, \text{TRUE})$$
$$= 0.0098 \text{ or } 0.98\% \tag{10.13}$$

The percentage of ball diameters above 20.00060 mm or $Z = +1.67$ can be calculated using MS Excel as follows:

$$1 - \text{NORMDIST}(20.00060, 20.00010, 0.00030, \text{TRUE})$$
$$= 0.0478 \text{ or } 4.78\% \tag{10.14}$$

During 100% sorting for ball diameter our estimated total rejects will be 0.98% + 4.78% = 5.76%. We have to produce and sort the following estimated amount of precision steel balls to fill our customer's order.

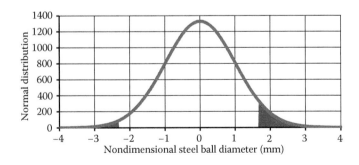

Figure 10.6 Precision steel ball diameter process distribution as a function of nondimensional steel ball diameter $Z = \dfrac{X - \bar{\bar{X}}}{\sigma}$. Customer's specification – 20.00000 ± 0.00060 mm is compared to our process capability as read on our gage. 0.98% of balls are rejected below $Z = -2.33$ and 4.78% of balls are rejected above $Z = +1.67$.

$$\text{Required production quantity for steel balls} = \frac{10,000}{(1-0.0576)} = 10,611$$

$$(10.15)$$

The result we obtained in Equation 10.5 is true if we are in ship-to-stock status to our customer, namely our laser beam gages are accepted as master gages for this order, or if our laser beam gages are perfectly correlated to our customer's receiving inspection gages.

Let us analyze another case in which we are not at ship-to-stock status to our customer and our customer's receiving inspection gages read ball diameters over the range of our customer's specification 0.00008 mm more than our laser beam gages; namely when I read a ball diameter of 20.00010 mm, my customer's receiving inspection reads 20.00018 mm. Our customer orders 10,000 precision balls with a diameter specification of 20.00000 ± 0.00060 mm to be measured with their receiving inspection gages. For our laser beam gages, the customer specification will shift to 19.99992 ± 0.00060 mm. Our production processes are in statistical control, ball diameters show a normal distribution shape, and the population average and the standard deviation for diameters of these particular steel balls as measured with our laser beam gages are given in Equations 10.11 and 10.12, respectively.

The red regions in Figure 10.7 represent ball diameter rejects from our process as read on our laser beam gages. The percentage of ball

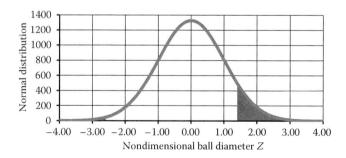

Figure 10.7 Precision steel ball diameter process distribution as a function of nondimensional steel ball diameter $Z = \dfrac{X - \bar{\bar{X}}}{\sigma}$. Customer's specification with a 0.00008 mm offset – 19.99992 ± 0.00060 mm is compared to our process capability as read on our gage. 0.466% of balls are rejected below $Z = -2.60$ and 8.076% of balls are rejected above $Z = +1.40$.

diameters below 19.99932 mm or Z = –2.60 can be calculated using MS Excel as follows:

$$\text{NORMDIST}(19.99932,20.00010,0.00030,\text{TRUE})$$
$$= 0.00466 \text{ or } 0.466\% \tag{10.16}$$

The percentage of ball diameters above 20.00052 mm or Z = +1.40 can be calculated using MS Excel as follows:

$$1 - \text{NORMDIST}(20.00052,20.00010,0.00030,\text{TRUE})$$
$$= 0.08076 \text{ or } 8.076\% \tag{10.17}$$

During 100% sorting for ball diameter our estimated total rejects will be 0.466% + 8.076% = 8.542%. We have to produce and sort the following estimated amount of precision steel balls in our factory to fill our customer's order per their receiving inspection requirements.

$$\text{Required production quantity for steel balls} = \frac{10,000}{(1-0.08542)} = 10,934 \tag{10.18}$$

Let us analyze one more case in which we have to deal with two independent critical dimensions of a product. The first critical dimension has a specification of 100.0 ± 10.0 μm. The second critical dimension has a specification of 25.0 ± 4.0 μm. Our production processes are in statistical control, both critical dimensions show a normal distribution shape, and population averages and standard deviations for these two critical dimensions are given below:

$$\bar{\bar{X}}_1 = 103.0 \ \mu m \quad \text{and} \quad \sigma_1 = 4.0 \ \mu m \tag{10.19}$$

and

$$\bar{\bar{X}}_2 = 24.0 \ \mu m \quad \text{and} \quad \sigma_2 = 2.5 \ \mu m \tag{10.20}$$

We sort our product 100% for these two independent dimensions, since our processes are not capable as shown in Figures 10.8 and 10.9.

Our customer ordered 100,000 of these products. How many products do we have to build and sort in order to fulfill our customer's order?

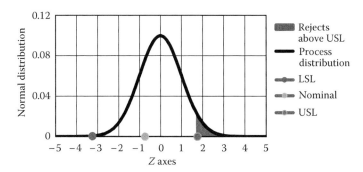

Figure 10.8 Dimension X_1 process distribution is shown along with lower, nominal, and upper specification limits as a function of Z.

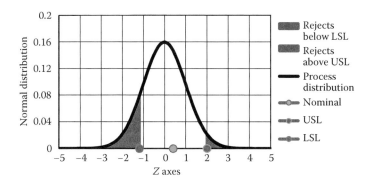

Figure 10.9 Dimension X_2 process distribution is shown along with lower, nominal, and upper specification limits as a function of Z.

Let us first find the cumulative probability of good parts for X_1. The percentage of rejects above the upper specification limit is

$$1 - \text{NORMDIST}(110,103,4,\text{TRUE})$$
$$= 0.040059 \text{ or } 4.0\% \qquad (10.21)$$

The percentage of rejects below the lower specification limit is

$$\text{NORMDIST}(90,103,4,\text{TRUE}) = 0.000577 \text{ or } 0.1\% \qquad (10.22)$$

The cumulative probability of good parts for X_1 is (100% − 4.0% − 0.1%) = 95.9%.

Now let us find the cumulative probability of good parts for X_2. The percentage of rejects above the upper specification limit is

$$1 - \text{NORMDIST}(29,24,2.5,\text{TRUE}) = 0.02275 \text{ or } 2.3\% \qquad (10.23)$$

The percentage of rejects below the lower specification limit is

$$\text{NORMDIST}(21,24,2.5,\text{TRUE}) = 0.11507 \text{ or } 11.5\% \quad (10.24)$$

The cumulative probability of good parts for X_2 is $(100\% - 2.3\% - 11.5\%) = 86.2\%$.

The probability of getting both X_1 and X_2 dimensions to be good on a part is the product of the preceding good probabilities, since the two dimensions are independent of each other.

$$\text{Probability that both } X_1 \text{ and } X_2 \text{ are good}$$
$$= (\text{Probability that } X_1 \text{ is good}) \times (\text{Probability that } X_2 \text{ is good})$$
$$= 0.959 \times 0.862 = 0.827 \text{ or } 82.7\% \quad (10.25)$$

The number of parts we have to produce and 100% sort to fulfill the 100,000 unit order is

$$\text{Number of parts to be built and 100\% sorted} = \frac{100,000}{0.827} = 120,898$$

$$(10.26)$$

Process capability (yields) make or break a company in high-volume production. Gages and processes that are under statistical control are the first laws of successful manufacturing structure. You might not have every process in your production to be 6σ capable, but the challenge is to get there with teamwork, pride, and determination.

Index

Page numbers with f and t refer to figures and tables, respectively.